ALTERNATING-CURRENT CIRCUITS
– Simply Explained!

…if you can solve a right-angled triangle, then you can easily solve **ANY** a.c. circuit <u>without memorizing **ANY** equations</u>!

Adrian Waygood, MEd (Tech)

First edition published 2018

©2018 Adrian Waygood

All rights reserved. No part of this book may be reprinted or reproduced or utilized in any form or by an electronic, mechanical, or other means, now known or hereafter invented, including photocopying and recording or in any information storage or retrieval system, without permission in writing from the author.

Table of Contents

Introduction 5
Chapter 1 *Introduction to Alternating Current* 7
Chapter 2 *Representing Sinusoidal Waveforms with Phasors* 25
Chapter 3 *Series Alternating-Current Circuits* 31
Chapter 4 *Energy and Power in Alternating Current Circuits* 71
Chapter 5 *Parallel Alternating-Current Circuits* 97
Where to Go From Here…? 131
Other Books by the Same Author 133

Introduction

In the 1960s, I was a student apprentice working for a regional electricity distribution company. At that time, student apprentices underwent what was then known as 'block release' training: alternately receiving on-the-job-training with my employer and studying at a local technical college.

I remember when we were first introduced to **alternating-current theory** at college. Our lecturer was obviously very knowledgeable about the subject but, unfortunately, had very little idea of how to *teach* it! He rapidly filled the chalkboard with a seemingly-endless range of equations as he talked, as we tried to understand what he was saying while, at the same time, desperately trying to scribble down everything he wrote on the chalkboard before he rubbed it off to make more room for even more equations! He told us, we needed to *'commit all these equations to memory'* if we were ever going to understand the topic! And, in those days, we never 'questioned' our lecturers. let alone ask him to slow down his delivery!

I'm afraid that I never was able to commit all those equations to memory and, consequently, never really fully-understood the behaviour of a.c. circuits! And I am ashamed to admit that I failed the final examinations and was put back a term. The second time around, however, I had a different lecturer who had a vastly-different approach to teaching alternating-current theory.

He told us that, while there were indeed a great many equations needed to understand alternating current, *none of them needed to be committed to memory!* He taught us how *every* a.c. circuits could be represented by an *electrical-equivalent of a vector diagram*, which he called 'phasor diagrams' (a newly-introduced term, at that time!), and *all* of these were, essentially, nothing more than **right-angles triangles**.

So, provided we could apply **Pythagoras's Theorem**, together with the **sine, cosine,** and **tangent trigonometric ratios**, to right-angled triangles, ALL those dreaded equations could be *derived* from these triangles *without having to commit **any** of them to memory!* And we could concentrate, instead, on really understanding how a.c. circuits behaved.

Thanks to this brilliant lecturer and his unique approach to teaching alternating-current theory, I not only flew through that term's finals, but went on to obtain my qualifications, and eventually become a lecturer myself.

I have been using the same approach, very successfully, to teach alternating-current theory to my students. And I present it to you, now, in this book.

I will make the same promise to *you*, as I have made to the hundreds of successful students I have taught over the years.

In order to understand the behaviour of any alternating-current circuit, single- or three-phase, you only need to be able to do the following:

    a.   *learn how to represent a circuit by means of a **phasor** (electrical vector) **diagram**.*

    b.   *commit to memory just **two** important equations.*

    c.   *remember and understand how to apply the mnemonic, **CIVIL**.*

    d.   *remember and apply **Pythagoras's Theorem** to right-angled triangles.*

    e.   *remember, and apply the **sine**, cosine, and **tangent** trigonometric ratios to right-angled triangles.*

*If you can do each of these, then you will be able to **generate ALL of the dozens of important equations, from scratch, without having to commit ANY of them to memory**!*

The book *does*, of course, assume that you have a basic understanding of the of d.c. theory: current, potential difference, resistance, energy and power, capacitance and inductance, and Ohm's Law.

The content of this book is based on the content of one of my other books, *'An Introduction to Electrical Science (2nd Edition)'* which, together with *'Electrical Science for Technicians'*, is available, internationally, from **Amazon** and other good book sellers.

If you like this particular book, then do please leave an honest review at Amazon.

Visit my blog, **www.professorelectron.com**, for additional resources and to leave any comments, suggestions, or corrections.

Adrian Waygood

October 2018

Chapter 1

Introduction to Alternating Current

On completion of this chapter, you must be able to

1. explain how the voltage induced into a conductor varies according to the angle at which the conductor moves through a magnetic field.

2. apply Fleming's Right-Hand Rule to determine the direction of the voltage induced into a conductor moving through a magnetic field.

3. explain why a conductor or loop, rotating within a magnetic field, generates a sinusoidal voltage.

4. explain each of the following terms specifying, where applicable, their SI units of measurement:
   a. amplitude
   b. instantaneous value
   c. period
   d. wavelength
   e. cycle
   f. frequency

5. given the peak value of an a.c. voltage or current, calculate its r.m.s. (or 'effective') value, and *vice versa*.

Introduction

In a metal conductor, free electrons are in a state of constant, random, and chaotic movement. When subjected to a constant electric field, this chaotic movement continues, but there is a tendency for these electrons to *drift* slowly (at a rate measured in just millimetres per hour!) along the conductor. This behaviour is illustrated in figure 1.1. We call this drift of free electrons an **'electric current'** which we express in **amperes** (symbol: A). When the net drift of free electrons is in just one direction, we call it **'direct current'** (**d.c.**).

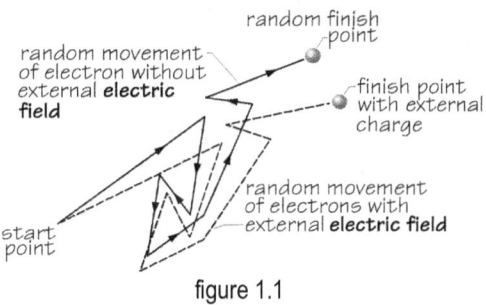

figure 1.1

When subjected to an electric field which periodically *reverses* direction, the chaotic movement of free electrons continues but, this time, there is a tendency for the net *drift* of those free electrons to also periodically reverse direction. We call this **'alternating current'** (a.c.).

> Although the abbreviation, **'a.c.'** stands for **'alternating current'**, we should be aware that it can also be used as an *adjective* to describe the word which follows. For example, it is perfectly correct to use the term, **'a.c. voltage'** to mean **'alternating voltage'**.

Generation of an a.c. voltage

Whenever there is relative movement between a conductor and a magnetic field, a potential difference (voltage) is *induced* into that conductor.

Faraday's Law tells us that the *magnitude* of this induced voltage is *directly proportional to the rate at which the lines of magnetic flux are cut by the conductor*. For a given velocity, the maximum rate will occur when the conductor cuts the flux perpendicularly (at right-angles); on the other hand, if the conductor runs *parallel* with the flux, then no flux is cut, and no voltage is induced into the conductor. This is illustrated in figure 1.2.

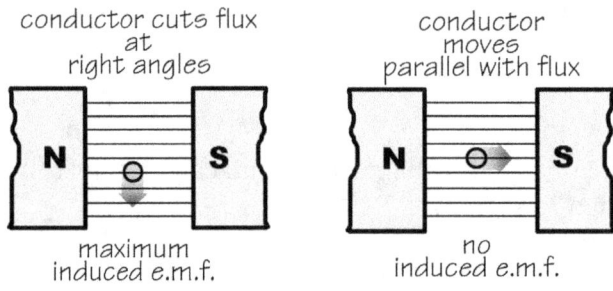

figure 1.2

The *rate* at which a conductor cuts lines of magnetic flux and, therefore, the value of the induced voltage, is determined by the *flux density* of the magnetic field, the *velocity* of the conductor as it moves through that field, and the *angle* at which the conductor cuts the lines of magnetic flux.

So, the general equation for this induced voltage is given by:

$$E = Blv \sin \theta$$

where:
- E = induced voltage (V)
- B = flux density (T)
- l = length of conductor (m)
- v = velocity of conductor (m/s)
- θ = angle cutting flux (°)

The *direction* of the induced voltage can be deduced by applying **Fleming's Right-Hand Rule** (for *conventional* current). For example, in figure 1.3, when the conductor moves vertically *downwards* through the magnetic field, Fleming's Right-

Hand Rule tells us that the positive end of the conductor is at the *nearest* end, and (conventional) current will emerge into the external circuit from that end, returning at the opposite end:

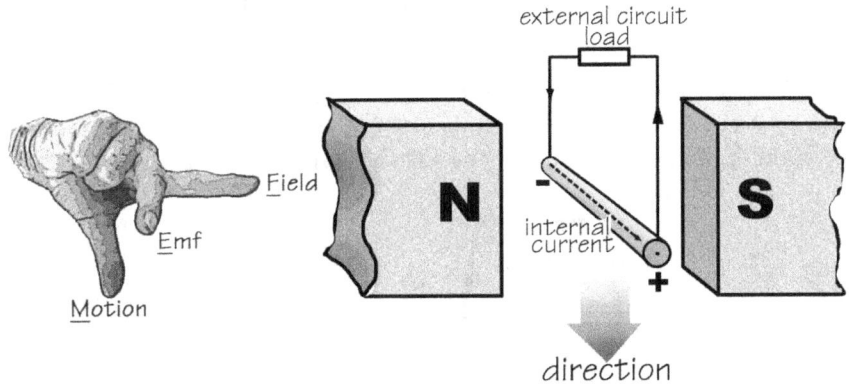

figure 1.3

A simple a.c. generator (alternator)

A simple **a.c. generator**, or **alternator** as it is known, consists of a single loop of wire (called an '**armature**'), pivoted so that it can rotate within the magnetic field set up between the poles of a magnet —as illustrated in figure 1.4.

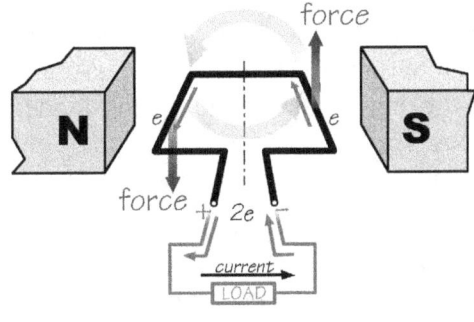

figure 1.4

When the loop is rotated by some external force provided by a **prime mover** (in the case of a 'real' alternator, a diesel engine, a steam turbine, a water turbine, a wind turbine, etc.), a potential difference is induced into each side of the loop. If we apply **Fleming's Right-Hand Rule** to each side of this loop, we will see that the potential difference induced into the left-hand side of the loop will act *towards* us, while the potential difference induced into the right-hand side of the loop will act *away* from us —as, by forming a loop, the two sides are in series with each other, these two induced voltages act to *reinforce* each other.

So if the voltage induced into each side of our loop is v volts, then the voltage induced into the complete loop must be $2v$ volts.

In a more-practical alternator, of course, the armature consists of a **coil**, *not* a single loop. So, for an armature winding having N turns, this voltage will be increased to

$2Nv$ volts. However, for the sake of simplicity, we'll continue to consider it as a single loop.

When this simple generator is connected to an external load, any resulting (conventional) load current will flow through the loop in the counter-clockwise direction shown in figure 1.4, above.

Connecting the rotating loop of our simple generator to its external load is achieved by brazing opposite ends of that loop to a pair of copper **slip rings** against which press spring-loaded blocks of carbon, called '**brushes**', as illustrated in figure 1.5:

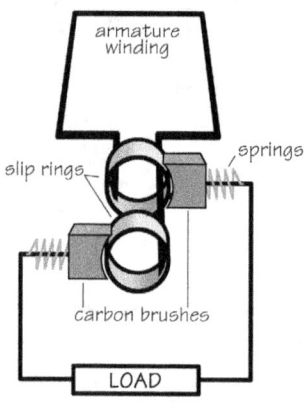

figure 1.5

Carbon is a reasonably-good conductor and has a *negative temperature-coefficient of resistance*, meaning that as its temperature *increases*, its resistance *decreases*. It is also *self-lubricating* so, as the slip rings rotate, the spring-loaded carbon brushes smoothly 'ride' on their surfaces, ensuring near-permanent contact between the rotating loop and the external circuit.

Just as the generator's loop rotates past the vertical position, the relative direction between each side of the coil and the magnetic flux reverses, causing the voltages induced into the coil to also reverse. So, the output-voltage from a generator equipped with slip rings periodically (twice for every revolution) reverses direction, producing (in theory at least) a *sinusoidal voltage waveform*. We will learn *why* this waveform is considered 'sinusoidal' in a moment.

To further increase the voltage induced into the winding, 'real' alternators, of course, use **electromagnets,** which are far more powerful than the permanent magnets shown in our simple model, and which significantly increases the flux density *(B)* of the magnetic field.

Generation of a sinusoidal voltage waveform

Let's now return to our simple model of an a.c. generator, to remind ourselves of how an a.c. voltage can be generated in a rotating armature loop. In the following series of diagrams (figures 1.6a – 1.6m), a conductor follows the counter-clockwise circular path, cutting the magnetic flux set up from the north towards the south magnetic poles. For the sake of simplicity, we'll consider just one side of the armature loop and we'll assume that, when that conductor cuts the flux at right-angles, a voltage of E_{max} volts is induced into the conductor:

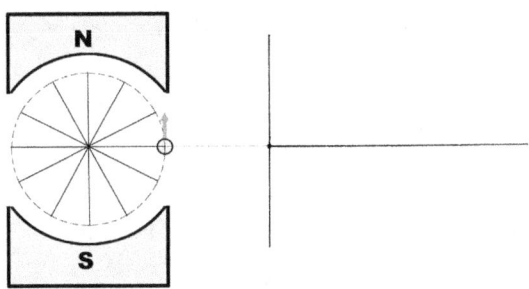

figure 1.6a

In this position, the conductor is moving parallel to the magnetic flux, so no voltage is induced into the conductor.

$$e = E_{max} \sin \theta = E_{max} \sin 0°$$
$$= 0 \text{ volts}$$

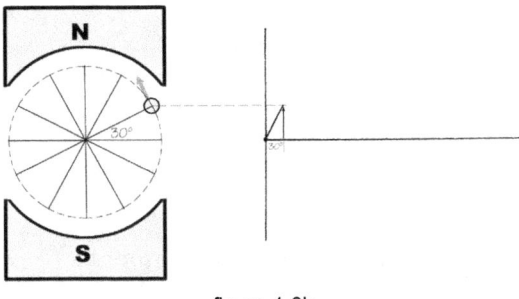

figure 1.6b

The conductor has moved along its circular path by 30°, and is cutting the flux by 30°, so a voltage is starting to be induced into the conductor.

$$e = E_{max} \sin \theta = E_{max} \sin 30°$$
$$= 0.5 \, E_{max} \text{ volts}$$

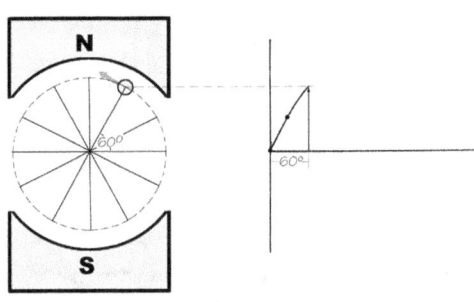

figure 1.6c

The conductor has now moved by along its circular path by 60°, and is now cutting the flux by 60°, and an even greater voltage is induced into the conductor.

$$e = E_{max} \sin \theta = E_{max} \sin 60°$$
$$= 0.866 \, E_{max} \text{ volts}$$

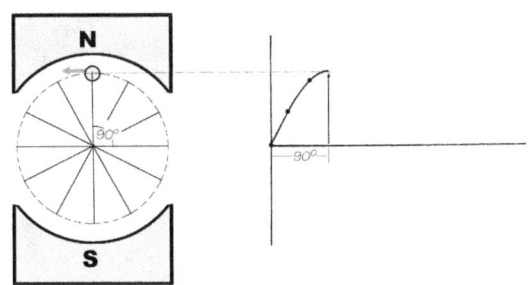

figure 1.6d

The conductor has now moved by along its circular path by 90°, and is now cutting the flux at right-angles, so the maximum voltage is induced into the conductor.

$$e = E_{max} \sin \theta = E_{max} \sin 90°$$
$$= 1.0 \, E_{max} \text{ volts}$$

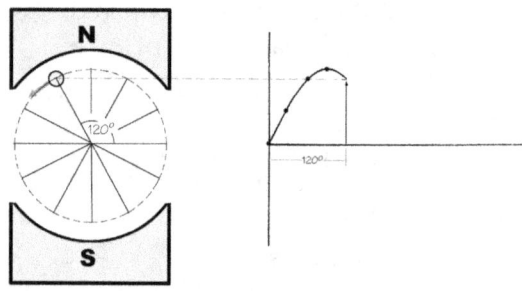

figure 1.6e

The conductor has now moved along its circular path by 120°, and is now cutting the flux by 60°, so the induced voltage is starting to fall.

$$e = E_{max} \sin \theta = E_{max} \sin 120°$$
$$= 0.866 \, E_{max} \text{ volts}$$

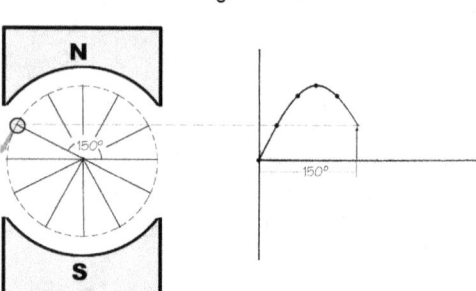

figure 1.6f

The conductor has now moved along its circular path by 150°, and is now cutting the flux by 30°, so the induced voltage has fallen further.

$$e = E_{max} \sin \theta = E_{max} \sin 150°$$
$$= 0.5 \, E_{max} \text{ volts}$$

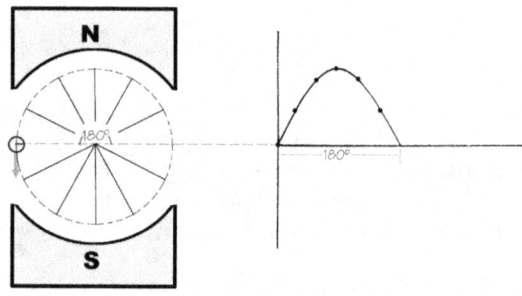

figure 1.6g

The conductor has now moved along its circular path by 180° and, once again, the conductor is moving parallel to the flux, so no voltage is induced into it.

$$e = E_{max} \sin \theta = E_{max} \sin 180°$$
$$= 0 \, E_{max} \text{ volts}$$

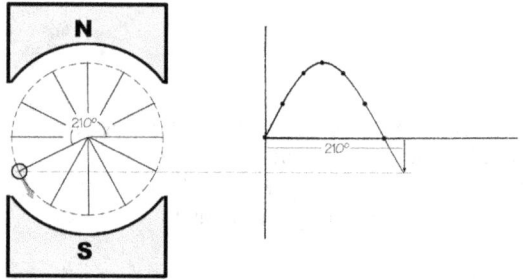

figure 1.6h

The conductor has now moved along its circular path by 210°, and the conductor is cutting the flux at 30° —but this time, in the opposite direction. So the induced voltage is now acting in the opposite sense.

$$e = E_{max} \sin \theta = E_{max} \sin 210°$$
$$= -0.5 \, E_{max} \text{ volts}$$

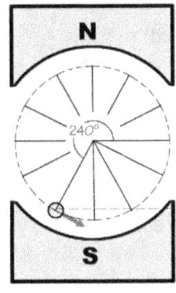
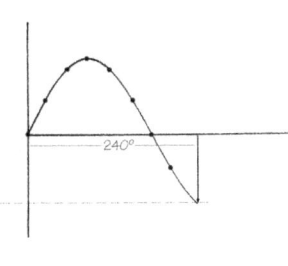

figure 1.6i

The conductor has now moved along its circular path by 240°, and the conductor is cutting the flux at 60°, so the induced voltage is now increasing in the opposite sense.

$$e = E_{max} \sin \theta = E_{max} \sin 240°$$
$$= -0.866 \, E_{max} \text{ volts}$$

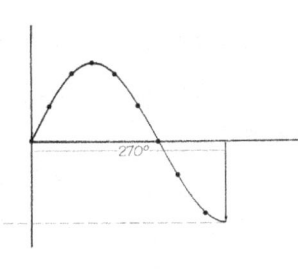

figure 1.6j

The conductor has now moved by along its circular path by 270°, and is now cutting the flux at right-angles (but in the opposite direction compared to the 90° position), so the maximum voltage is induced into the conductor.

$$e = E_{max} \sin \theta = E_{max} \sin 270°$$
$$= -1.00 \, E_{max} \text{ volts}$$

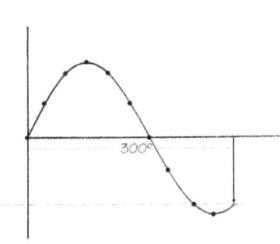

figure 1.6k

The conductor has now moved by along its circular path by 300°, and is now cutting the flux at 60°, so the induced voltage is starting to fall again.

$$e = E_{max} \sin \theta = E_{max} \sin 300°$$
$$= -0.866 \, E_{max} \text{ volts}$$

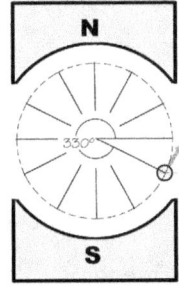
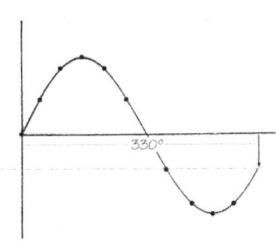

figure 1.6l

The conductor has now moved by along its circular path by 330°, and is now cutting the flux at 30°, so the induced voltage is starting to fall further.

$$e = E_{max} \sin \theta = E_{max} \sin 330°$$
$$= -0.5 \, E_{max} \text{ volts}$$

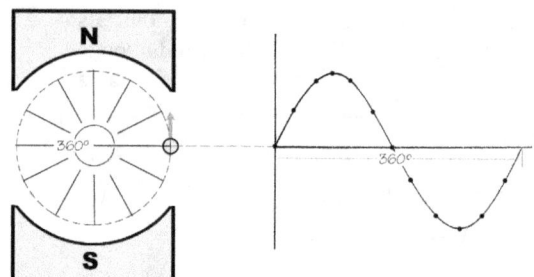

The conductor has now moved through 360° and is, once again, moving parallel to the flux and no voltage is induced into the conductor.

$$e = E_{max} \sin \theta = E_{max} \sin 360°$$
$$= 0\ E_{max} \text{ volts}$$

figure 1.6m

So, as the conductor (or, in practice, a **loop**) moves through its circular path between the magnetic poles, the value of the voltage induced into it continually varies from zero, to some maximum value (E_{max}) in one direction, then back to zero and, finally, to the same maximum value but in the *opposite* direction.

If we were to now draw the waveform to scale, and check the values of the instantaneous voltages at, say, 30° intervals from 0° to 360°, figure 1.7 would be the result:

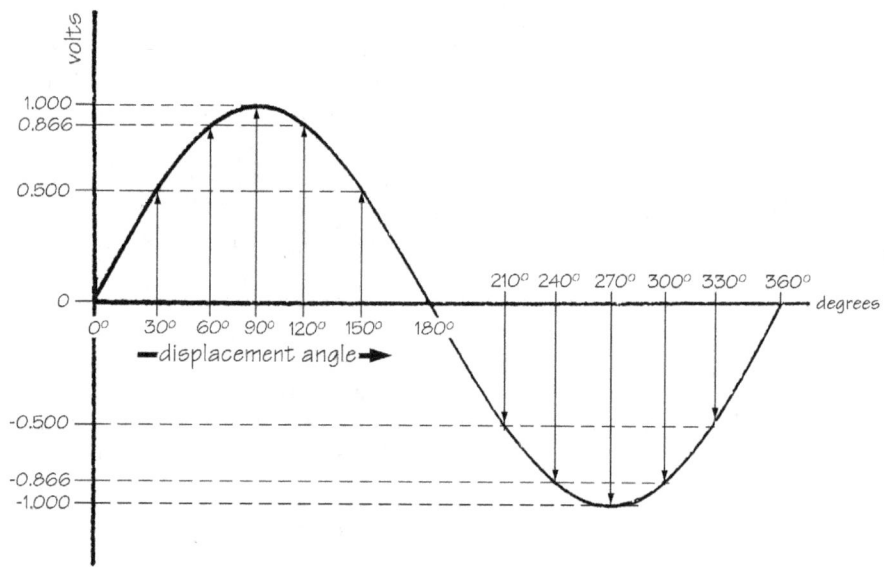

figure 1.7

The angle, expressed in 'electrical' degrees, measured from the origin, at which each of these instantaneous values of voltage occur, is termed their **displacement angle** (symbol: θ, pronounced 'theta'). In general terms, then, the values of each of these instantaneous voltages *(e)* correspond to the peak voltage (E_{max}) multiplied by the *sine* of corresponding displacement angle *(θ)*, i.e:

instantaneous voltage = peak voltage × sine of displacement angle

$$e = E_{max} \sin \theta$$

Because each instantaneous voltage is proportional to the *sine* of the displacement angle, the entire generated waveform is called a '**sine wave**'.

So, the **instantaneous voltage** (symbol: *e*) at any point along the sine wave, then, is given by the equation:

$$e = E_{max} \sin \theta$$

where: *e* = instananeous voltage
E_{max} = peak voltage
θ = displacement angle

Since current is directly proportional to voltage, so we can also say:

$$i = I_{max} \sin \theta$$

where: *i* = instananeous current
I_{max} = peak current
θ = displacement angle

Note the use of the terms, '**electrical degrees**', earlier. In the case of a simple, *two-pole alternator*, like the one shown above, *one* complete revolution (i.e. 360 *mechanical* or *physical* degrees) of the armature results in *one* complete cycle comprising 360 *electrical* degrees. But, 'real' alternators are practically always **multi-pole** machines. In the case of a *four*-pole machine, for example, *one* complete revolution (360 *mechanical* or *physical* degrees) by the armature results in an output of *two* complete voltage cycles —i.e. 720 *electrical* degrees.

So, depending on the number of poles, there is a *difference* between the number of 'mechanical' degrees physically rotated by the loop and the number of resulting 'electrical' degrees,. Whenever we describe waveforms, therefore, we should realize that *we always talk in terms of 'electrical degrees'*.

Terminology used to describe a.c. waveforms

Figure 1.8 summarises the **terminology** we use to describe the various parts of any sine wave (voltage or current):

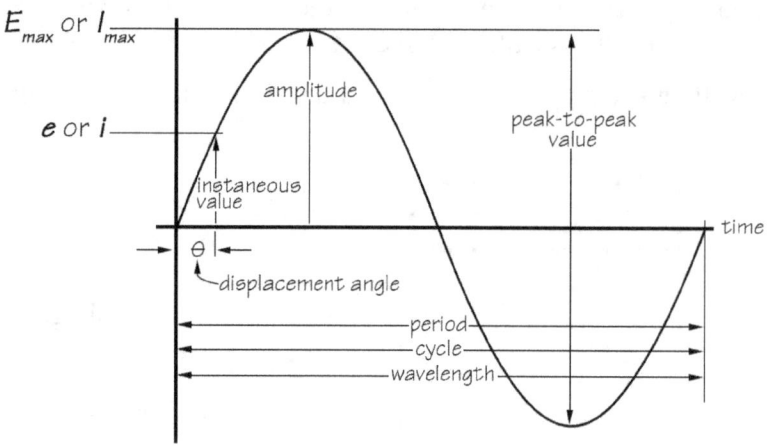

figure 1.8

Amplitude:	The peak value of an a.c. waveform, in either the positive *or* in the negative sense, symbols: E_{max} or I_{max}, measured in volts or amperes.
Peak-to-peak value:	Twice the amplitude of an a.c. waveform. Symbols: E_{p-p} or I_{p-p}.
Instantaneous value:	The value of voltage or current, at any specified displacement angle, during a complete cycle, symbols: *e* or *i*, measured in volts or amperes.
Displacement angle:	The displacement, symbol: *θ* (the Greek letter, '*theta*'), of any instantaneous value of voltage or current, expressed in electrical degrees and measured from the origin of the sine wave..
Period, or **Periodic time:**	The time taken to complete one complete cycle, symbol: *T*, measured in seconds.
Wavelength:	The distance between two displacements of the same phase, symbol: λ (the Greek letter, '*lambda*'), measured in metres.
Cycle:	One complete set of changes in the values of a recurring variable quantity, such as voltage or current.
Frequency:	The number of complete cycles completed per unit time, symbol: *f*, measured in hertz (Hz). It's also the reciprocal of the periodic time.

How we measure a.c. sinusoidal values

Assigning values to direct voltages or currents is quite straightforward, because they are constant values. But the value of a sinusoidal voltage or current is continuously changing in both *magnitude* and *direction*, so how can we possibly assign any meaningful value to it?

We *could*, of course, simply specify its *peak value* or *amplitude* (E_{max} or I_{max}) —but, as the waveform only reaches this value *twice* during any complete cycle, it is hardly representative of the entire waveform.

So, why not use its *average value*, instead? Well, if we were to work out the waveform's average value over a complete cycle, it will work out at *zero* —because, of course, the average value over the *positive* half-cycle will be completely cancelled by its average value over the *negative* half-cycle!

If neither the **peak value**, nor the **average value**, represent meaningful ways of expressing a sinusoidal voltage or current over a complete cycle, then *how do we proceed?*

Well, let's consider the following analogy (figure 1.11).

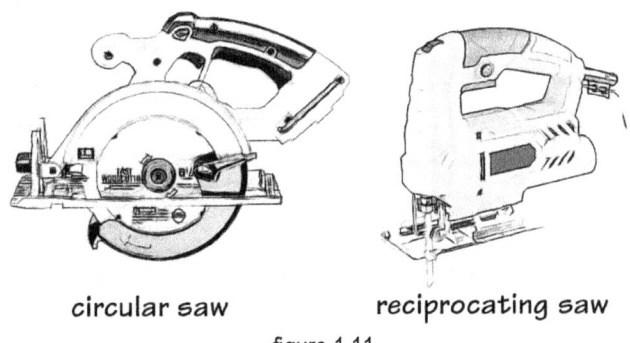

circular saw   reciprocating saw

figure 1.11

Let's think, for a moment, how we might compare the operating speed of a **reciprocating jigsaw** to that of a **circular saw**. The jig-saw blade, of course, has *linear* motion (it is continuously oscillating up and down), whereas the circular-saw blade is continuously *rotating* —so it's rather difficult to compare their 'speeds' as they each operate in completely-different ways.

However, what if we were to compare them, instead, in terms of *the rate at which they each cut through timber?* In other words, what if we were to say that if the jig-saw cuts through a given length of timber at *exactly* the same rate as the circular saw, then its 'reciprocating speed' must be *'equivalent'* to the 'rotational speed' of the circular saw?

Well, we apply a similar idea when we measure a.c. However, in this case, *we compare the **work** an a.c. current does with the **work** that a d.c. current does.*

We know that if a voltage is applied to a resistive circuit, the resulting current will cause the temperature of the resistor to rise, and this will happen *regardless* of whether the current is *direct* current or *alternating* current, so we make use of this property.

Let's consider the simple experiment, shown in figure 1.10.

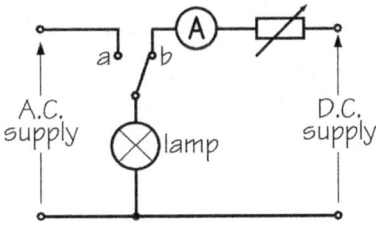

figure 1.10

With the switch in position *a*, an incandescent lamp is connected to an a.c. supply, and the brightness (or, more correctly, 'luminous intensity') of the lamp is carefully measured, under laboratory conditions, using a photoelectric meter. Next, the switch is moved to position *b*, and the variable resistor is adjusted until *exactly the same brightness is achieved*. The ammeter now indicates the value of d.c. current that has produced *exactly the same brightness* (in other words, produces *exactly the same heating effect*) as that produced by the a.c. current. So if the ammeter indicates a direct current of, say, 0.5 A when identical luminous intensity is achieved, then the 'effective value' of the a.c. current is also considered to be 0.5 A.

> So, we can say that *'the **effective value** of an alternating current is measured in terms of a direct current which produces exactly the same heating effect in the same resistance'*.

If we were to compare the a.c. current's **effective value** to its **peak value**, using an oscilloscope (an instrument that allows us to observe, and to measure, waveforms), we would find that, for a sine wave, its effective value is equal to **0.707** times its peak value. Non-sinusoidal waveforms would produce different effective values.

So, the relationship between the effective value of this a.c. current and its peak value would be as illustrated in figure 1.11:

$$I_{effective} = 0.707 \times I_{max}$$

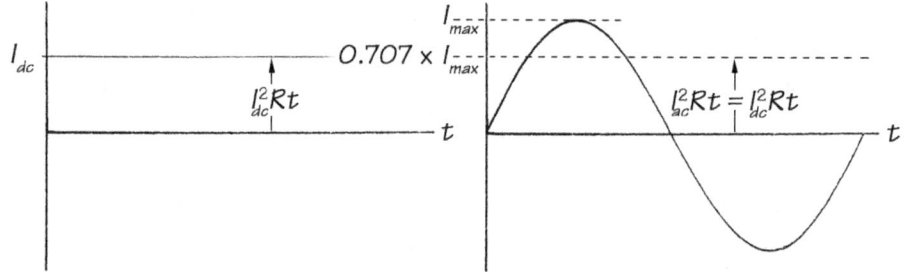

for a.c., 0.707 I_{max} does exactly the same work as I_{dc}.

figure 1.11

Since voltage and current are directly proportional to each other, we can also say:

$$E_{effective} = 0.707 \times E_{max}$$

The '**effective value**' of an a.c. current or voltage is more-commonly known as its '**root-mean-square**' or '**r.m.s. value**'. The expression, '**root-mean-square**', is derived from the mathematical proof for obtaining this value of 0.707. We won't to go into this proof in any great detail as it's well beyond the scope of this book, but we should at least be aware of the *origin* of this term.

In the above explanation of how we determine the effective value, we said it is based on the work *(W)* done by a current *(I)* passing through a resistance *(R)* over a particular period of time *(t)*. We know that the equation for this is: $W = I^2Rt$. So, we start by dividing the a.c. current waveform up into a great many (the *more* the better!) *instantaneous values of current:* i_1, i_2, i_3, i_4, etc., as shown in figure 1.12:

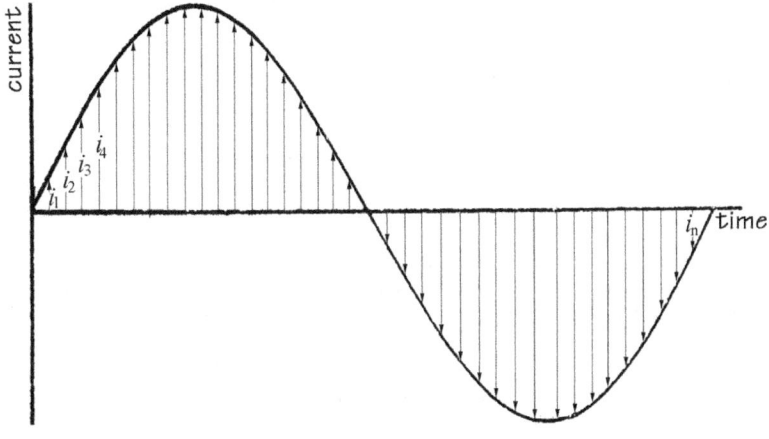

figure 1.12

If we then apply the above equation to *each and every one* of these instantaneous currents, we'll end up with the following equation:

$$I^2_{effective} R t = i_1^2 R t + i_2^2 R t + i_3^2 R t + i_4^2 R t + etc. \text{ (for the whole waveform)}$$

By dividing throughout by $R\,t$ we can eliminate both R and t (both being constants, of course) from the equation:

$$I^2_{effective} \frac{\cancel{R}\cancel{t}}{\cancel{R}\cancel{t}} = i_1^2 \frac{\cancel{R}\cancel{t}}{\cancel{R}\cancel{t}} + i_2^2 \frac{\cancel{R}\cancel{t}}{\cancel{R}\cancel{t}} + i_3^2 \frac{\cancel{R}\cancel{t}}{\cancel{R}\cancel{t}} + i_4^2 \frac{\cancel{R}\cancel{t}}{\cancel{R}\cancel{t}} + etc. \text{ (for the whole waveform)}$$

$$I^2_{effective} = i_1^2 + i_2^2 + i_3^2 + i_4^2 + etc. \text{ (for the whole waveform)}$$

Next, we find the **mean** (average) for *all* those individual **squared** instantaneous currents:

$$I^2_{effective} = \frac{i_1^2 + i_2^2 + i_3^2 + i_4^2 + etc. \text{ (for the whole waveform)}}{n}$$

(where *n* represents the number of individual instantaneous currents)

Finally, to eliminate the squares, we can find the **square-root** of both sides of the equation:

$$I_{effective} = \sqrt{\frac{i_1^2 + i_2^2 + i_3^2 + i_4^2 + etc. \text{ (for the whole waveform)}}{n}}$$

So, as we can now see, the **effective value of current** is equal to the **square-root** of the **mean** (i.e. 'average') of the **squares** of each of the individual instantaneous currents. —hence, the term: **'root-mean-square'**. If we were to insert *actual* values of all those individual instantaneous currents into the above equation then, for a sine-wave, it would *always* work out to be **0.707 I_{max}**.

To obtain an accurate result, in practice this calculation is usually performed using calculus, rather than by the technique described above.

This figure of 0.707 would, of course, not apply to *non-sinusoidal waveforms*, such as square- or triangular-shaped waveforms that we may come across in electronic circuits. But this is beyond the scope of this book, as we will only be working with sine waves.

To summarise:

> The **effective**, or **r.m.s. value** of a sinusoidal current is given by:
> $$I_{rms} = 0.707\, I_{max}$$

And, since voltage and current are proportional to each other:

> The **effective**, or **r.m.s. value** of a sinusoidal voltage is given by:
> $$E_{rms} = 0.707\, E_{max}$$

It's important to understand that a.c. currents and voltages are *always* expressed in **r.m.s. values** *unless otherwise specified*. For example, *all* a.c. ammeters and voltmeters are calibrated to indicate r.m.s. values. So, we do *not* normally need to add the subscript 'rms' to the symbols for current or voltage, other than for those occasions when clarification is necessary —in other words, we usually simply write 'I' or 'E', rather than 'I_{rms}' or 'E_{rms}'.

> **Important!**
> Unless otherwise specified, **a.c. values are <u>*always*</u>** quoted in **r.m.s. values**, and *all* a.c. ammeters and voltmeters are calibrated to provide readings in **r.m.s. values**. Because of this, we will rarely see the subscript 'rms' used, except where clarification is necessary.

Worked Example 1

What is the peak value of the nominal 230-V a.c. European residential supply?

Solution

(For clarification, we'll retain the subscript 'rms' in this example.)

As *all* a.c. values are normally expressed as r.m.s. values, then 230 V is an r.m.s. value, so:

$$\text{since} \quad E_{rms} = 0.707 \, E_{max}$$

$$\text{then} \quad E_{max} = \frac{E_{rms}}{0.707} = \frac{230}{0.707} = 325 \text{ V (Answer)}$$

Worked Example 2

What is the peak value of the nominal 120-V North-American residential supply?

Solution

(For clarification, we'll retain the subscript 'rms' in this example.)

Since *all* a.c. values are normally expressed as r.m.s. values, then 120 V is an r.m.s. value, so:

$$\text{since} \quad E_{rms} = 0.707 \, E_{max}$$

$$\text{then} \quad E_{max} = \frac{E_{rms}}{0.707} = \frac{120}{0.707} = 170 \text{ V (Answer)}$$

Worked Example 3

Using an oscilloscope, we measure the peak value of a sine-wave voltage across a resistor as being 250 mV. What is its r.m.s. value?

Solution

(Again, for clarification, we'll retain the 'rms' subscript in this example.)

$$E_{rms} = 0.707 \, E_{max}$$
$$= 0.707 \times (250 \times 10^{-3}) = 177 \times 10^{-3} \text{ V} \quad \text{or} \quad 177 \text{ mV (Answer)}$$

Review what you have learnt

Now that you have completed this chapter, go back and look at the **objectives** listed at the beginning. If you place a question mark at the end of each objective, and ask yourself, *'Can I...'*, then those objectives become **test items**. If you can answer each test item correctly, then you can move on to the next chapter.

Chapter 2

Introduction to phasors

On completion of this chapter, you must be able to

1. describe the differences and similarities between an electrical 'phasor' and a mechanical 'vector'.
2. specify what constitutes a 'positive' angle, when measuring the position of a phasor.
3. explain how we add two phasors that respresent the same quantity.

Representing sinusoidal waveforms with phasors

Most of us are already familiar with the concept of simple **vectors**.

Quantities such as **force** or **velocity** can be represented by means of a **vector** —where the *length* of that vector represents the *magnitude* of the force (in newtons) or the velocity (in metres per second), while the *direction* in which that vector points represents the *direction* in which the force or velocity is acting. Two or more forces acting on the same object can be added or subtracted, by simply adding or subtracting their corresponding vectors —this can be done either graphically, for example by drawing a '**triangle of forces**' to scale, or by applying the rules of simple geometry and trigonometry. The same, of course, applies to velocities.

'**Phasors**' are, to electrical engineering, what '**vectors**' are to mechanical engineering. In fact, the term 'phasor' didn't even appear until the 1960s; prior to that, they were known simply as '**electrical vectors**' and, for the reason described below, they are still often referred to as '**rotating vectors**'.

But there are important differences between phasors and vectors.

For example, while the *length* of a phasor, just like a vector, represents the *magnitude* of an alternating-voltage or current, its *'direction'* (or, more accurately, its *'angle'*) *doesn't* represent 'direction' but, rather, the *time displacement* of that voltage or current —expressed in terms of *displacement angle measured in a counter-clockwise direction*.

This is illustrated in figure 2.1.

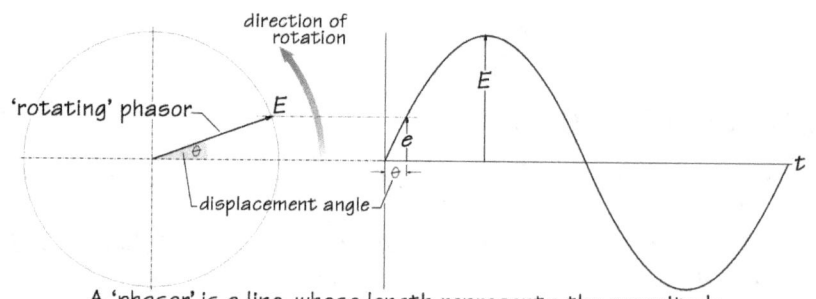

A 'phasor' is a line, whose length represents the magnitude of an a.c. voltage or current, and whose angle, measured counterclockwise from the real positive axis, represents its displacement angle.

figure 2.1

In figure 2.1, the **length** of the phasor, E, is equal to the **peak-value** or **amplitude** of the voltage. Its vertical component, e, represents the corresponding **instantaneous value** of voltage for any given **displacement angle**, θ, measured counter-clockwise.

If we can imagine the phasor, E, rotating counter-clockwise, then its corresponding instantaneous value, e, projected to the right, will vary in value from 0 V to $\pm E_{max}$, as it traces out a complete sinusoidal waveform every 360 electrical degrees.

> At this point, you may wish to search the world-wide web for *'electrical phasors'*, as there are numerous sites, such as *YouTube*, which use short video clips to demonstrate how phasors or 'rotating vectors' trace out sinusoidal waves far more clearly than can ever be described in a book.

The phasor's **arrow-head** serves *two* useful functions: (1) it identifies the 'rotating' end (as opposed to the 'fixed' end) of the phasor, and (2) it helps distinguish two or more phasors from each other, should they each lie at the same angle (i.e. superimposed one over the other).

It cannot be over-emphasized that the **displacement angle** *(θ)* is *always* measured, **counter-clockwise** from the real (horizontal) positive axis.

The importance of phasors is that they can be used to enable us to *vectorially-add* or to *vectorially-subtract* two or more voltage (or two or more current) waveforms which are out of phase with each other. Phasors also enable us to determine the *phase-relationship* between a voltage waveform and a current waveform in an a.c. circuit —that is, to shown whether one waveform *leads* or *lags* the other.

Although phasors are considered to be continuously 'rotating' in a counter-clockwise direction, in order to *analyse* them, we need to 'freeze' them in time. As we shall learn shortly, we usually do this when one or other of the phasor quantities is lying along the horizontal positive ('real') axis.

Phasors that represent the **same quantities** (i.e. *either* voltages *or* currents), but whose waveforms are out of phase with each other, may be added or subtracted in *exactly* the same way as vectors are added or subtracted, and by using *exactly* the same techniques —graphically (just like the 'triangle of forces') or mathematically—as illustrated in figure 2.2.

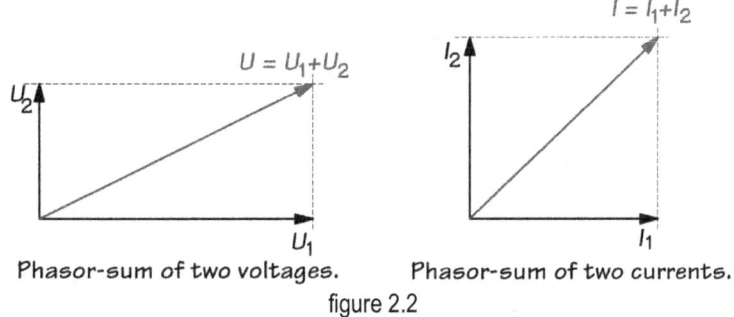

Phasor-sum of two voltages.   Phasor-sum of two currents.

figure 2.2

For example, in the case of the two a.c. voltage drops, U_1 and U_2, we can add them *vectorially* to find the total voltage drop, U. Similarly, in the case of two branch currents, I_1 and I_2, we can add them *vectorially* to find their total, I.

Phasors which represent **different quantities** (a voltage *and* a current) obviously *cannot be added or subtracted*, however the *angle* between them is important because it represents the phase relationship, or 'phase angle' (ø —pronounced *'phi'*), between those two quantities, as illustrated in figure 2.3. Remember, phase angles are always measured in a *counter-clockwise* direction.

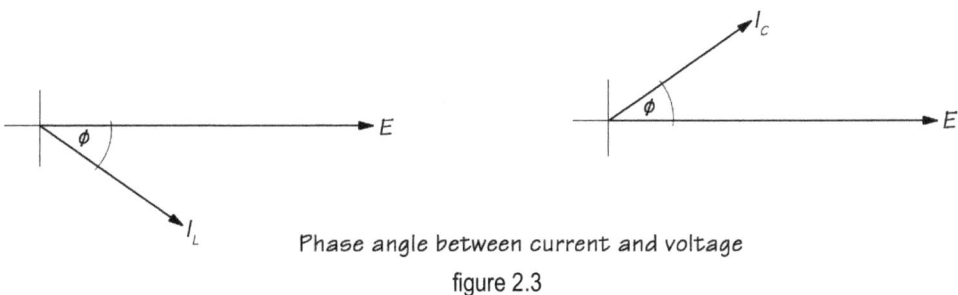

Phase angle between current and voltage

figure 2.3

If you imagine the two phasors, shown in figure 2.3, rotating in a counter-clockwise direction, the current, I_L, will *lag behind* its voltage, E, whereas the current, I_C, will *lead* its voltage, E.

So, in these examples, we say that the current, I_L, is **'lagging'** the voltage, whereas the current, I_C, is described as **'leading'** the voltage. The angle by which a current either lags, or leads, its voltage is termed the **'phase angle'** for the circuit which the phasor diagram represents.

So, the term, **'lagging'**, describes a phasor that lies counter-clockwise relative to a reference phasor (for example, E in figure 2.3). The term, **'leading'**, describes a phasor that lies clockwise relative to a reference phasor.

Although the *length* of a phasor, strictly, represents the maximum or *peak value* of a voltage or current, when we come to actually use them to solve a.c. circuits, we always allocate **r.m.s. values** to them. This is perfectly legitimate because an r.m.s.

value is directly proportional to a peak value, and whenever we work with alternating quantities, they are *always* expressed in r.m.s. values, *never* peak values.

> A **phasor** is an arrowed line whose *length* represents the *magnitude* of an alternating voltage or current and whose *angle*, measured counter-clockwise from the horizontal positive ('real') axis, represents the *displacement angle* of that voltage or current.
>
> A phasor's arrow-head represents the 'rotating' end of that phasor and *not* its direction.
>
> The rules of **vector addition** and **subtraction** apply to phasors representing like quantities (i.e. two or more voltage phasors, or two or more current phasors).

As we shall soon see, *phasors are, thankfully, very much easier to **use** than they are to **explain**!* So, if you haven't yet quite got your head around 'phasors', don't worry! Things will become clearer when we start to actually use them…

Table 2.4, below, summarises what we have learnt about phasors.

PHASORS
- the <u>length</u> of a phasor represents a waveform's amplitude.
- the <u>angle</u> of a phasor represents the time-angle between a waveform and a second (reference) waveform. Angles are measured counterclockwise.

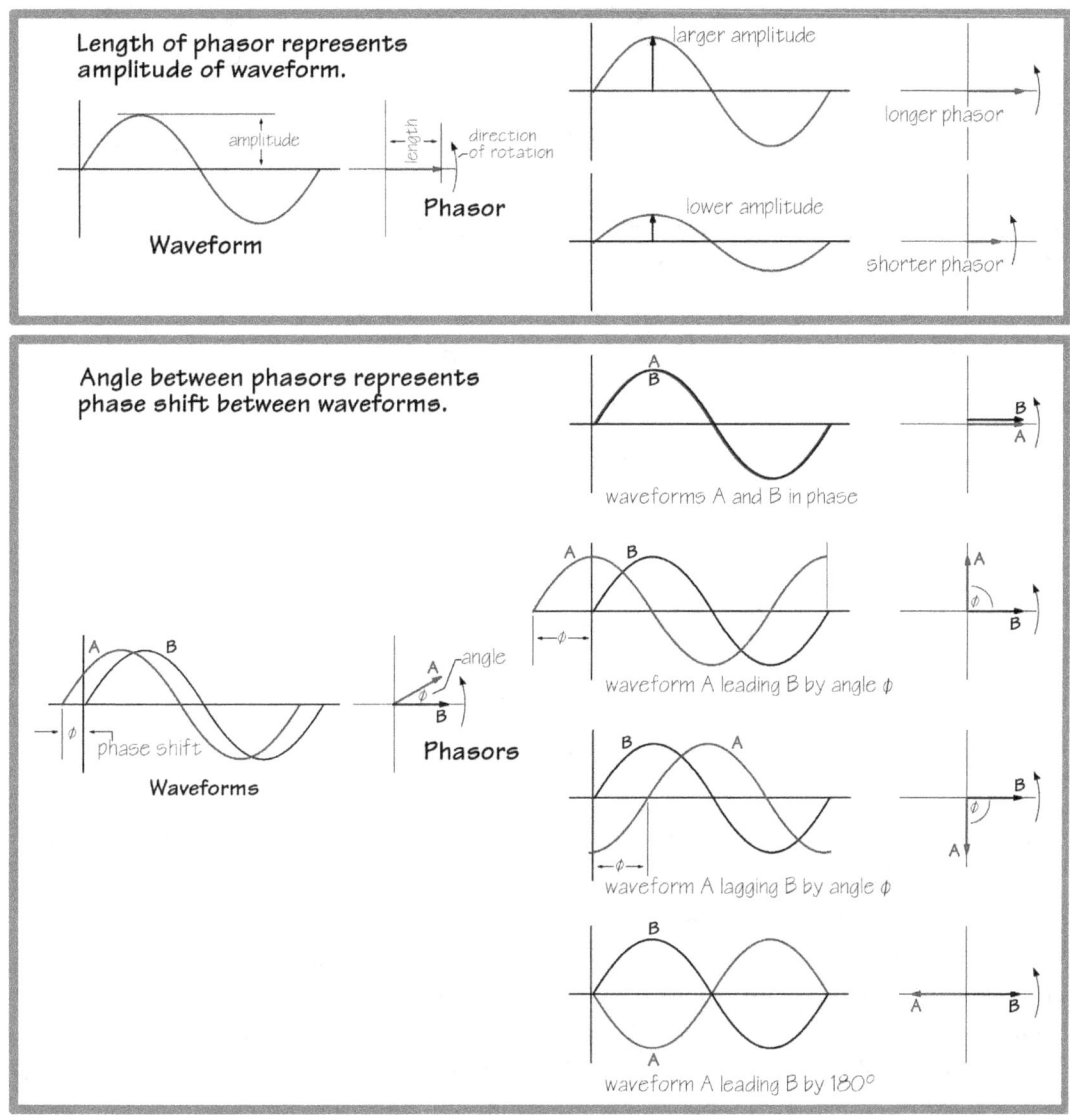

table 2.4

All you need to know to understand alternating current

Understanding phasors is the **key** to understanding and solving a.c. circuits. Period!

The chapters that follow explain *how* phasors can be used to understand, and to solve problems on, the behaviour of alternating-current circuits.

If you

- can represent an a.c. circuit with a phasor diagram,

- can add or subtract two vectors,
- understand, and can apply, Pythagoras's Theorem, and
- can apply the *sine*, *cosine,* and *tangent* trigonometric ratio (in particular, the *cosine* ratio), to simple right-angled triangles…

…then you are already *well on your way to being able to solve most problems* relating to behaviour of **series and parallel a.c. circuits**, and you will also be in a good position to understand and solve **series-parallel single-phase a.c. circuits** as well as **three-phase ac circuits**!

And you will be able to do *all* of this ***without having to remember more than just two of the many equations associated with a.c. theory****!*

Let's summarise on the promise made in the previous paragraph.

To understand the behaviour of a.c. circuits, you will need to:

- recall, and be able to apply, **Pythagoras's Theorem**.
- recall, and be able to apply to right-angled triangles, each of the following **trigonometric ratios**:

$$\sin \phi = \frac{\text{opposite}}{\text{hypotenuse}}; \quad \cos \phi = \frac{\text{adjacent}}{\text{hypotenuse}}; \quad \tan \phi = \frac{\text{opposite}}{\text{adjacent}}$$

- memorize, and be able to apply, the following two equations:

$$X_L = 2\pi f L \quad \text{and} \quad X_C = \frac{1}{2\pi f C}$$

- memorize, and be able to apply, the mnemonic **CIVIL** —but more of this, later.

Really! That's *all* there is to it! Well… not quite! Of *course*, there really *is* more to a.c. theory than just that —but, armed with only the knowledge described above, you will soon be well on your way to gaining a tremendous understanding of the behaviour of a single-phase a.c. circuits, and you will quickly be able to solve a great many of the problems that you will be faced with both in the classroom and on the job.

All phasor diagrams are basically just **right-angled triangles**. So, *'solving an a.c. problem'* amounts to little more than *'solving a right-angled triangle problem'*: i.e. finding the lengths of its sides and the angles between them.

Review what you have learnt

Now that you have completed this chapter, go back and look at the **objectives** listed at the beginning. If you place a question mark at the end of each objective, and ask yourself, *'Can I...'*, then those objectives become **test items**. If you can answer each test item correctly, then you can move on to the next chapter.

Chapter 3

Series Alternating-Current Circuits

On completion of this chapter, you must be able to

1. sketch a waveform for each of the following, showing the phase-relationship between the current and supply-voltage:

    a. purely-resistive circuit.

    b. purely-inductive circuit.

    c. purely-capacitive circuit.

    d. series *R-L* circuit.

    e. series *R-C* circuit.

2. state the phase-relationship between the current and supply-voltage for

    a. a purely-resistive circuit.

    b. a purely-inductive circuit.

    c. a purely-capacitive circuit.

3. sketch the phasor diagram representing a

    a. purely-resistive circuit.

    b. purely-inductive circuit.

    c. purely-capacitive circuit.

    d. series *R-L* circuit.

    e. series *R-C* circuit.

    f. series *R-L-C* circuit.

4. develop an impedance diagram for a

    a. series *R-L* circuit.

    b. series *R-C* circuit.

    c. series *R-L-C* circuit.

5. from impedance diagrams, derive equations for resistance, inductive reactance, capacitive reactance, and impedance, in terms of voltages and currents.

6. state the equation for inductive reactance, in terms of inductance and frequency.

7. state the equation for capacitive reactance, in terms of capacitance and frequency.

8.   explain what is meant by the term, 'series resonance'.

9.   list the effects of series resonance.

10.   solve problems on series a.c. circuits, including series-resonant circuits.

Introduction

> **Important!** The key to understanding and solving a.c. circuits, whether they are series circuit, parallel circuits, or even three-phase circuits, is your ability to *sketch a phasor diagram (a triangle)* which represents that circuit and, then, to use this phasor diagram *to generate all the equations you need*, simply *by treating it as a right-angled triangle, and applying simple geometry or trigonometry to solve it*.

All practical alternating-current circuits exhibit combinations of **resistance** (symbol: R), **inductance** (symbol: L), and **capacitance** (symbol: C). The amount of each of these quantities appearing in any particular circuit is determined by the *configuration* and *design-characteristics* of that circuit.

For example, *all* conductors (by virtue of their length, cross-sectional area, and resistivity) exhibit *natural* amounts of resistance. Overhead power lines, due to the configuration of their individual conductors, exhibit relatively high *natural* values of inductance, as well as some capacitance and resistance. Underground cables and even residential-wiring cables, because of the closeness of their individual conductors, exhibit relatively high *natural* values of capacitance as, well as some inductance and resistance.

But we can also *modify* the natural resistance, inductance, and capacitance of any circuit by adding **resistors**, **inductors**, or **capacitors**.

'Real' a.c. circuits, then, are relatively complicated because they contain a *combination* of resistance, inductance, and capacitance. So, in order to understand the behaviour of any a.c. circuit, it is first necessary to start by considering how an 'ideal' circuit would behave if it existed.

In this context, an 'ideal' circuit is one that is '**purely resistive**', '**purely inductive**', or '**purely capacitive**'.

'Ideal' circuits *only exist in theory*. But if we are able to understand how these, relatively-simple, ideal circuits *would* behave if they *did* exist, then we will be able to move on to *combine* these individual behaviours in order to understand how *real* and *more complicated* circuits behave.

> **Important!**
> In the circuit diagrams that follow, it is important to understand that the symbols represent quantities *not* components. That is, **resistance** *not* resistors; **inductance** *not* inductors; **capacitance** *not* capacitors.

Throughout the rest of this chapter, the voltages and currents we will be referring to are '**phasor**' (vectorial) quantities. So, in order to remind ourselves of this, the symbols for voltage and current will be shown with small 'bars' placed above them (i.e. $\overline{E}, \overline{U}_R, \overline{U}_L, \overline{U}_C,$ and $\overline{I}$). This is one recognized way (another is by **bolding** the symbols) of indicating that these are phasor quantities, rather than scalar quantities. It's unncessary to do this when labelling phasor diagrams as the quantities involved are obviously phasors.

The *opposition* to the passage of a.c. current (i.e. resistance, inductive reactance, capacitive reactance, and impedance) are *not* phasor quantities and, so, should *not* have bars placed above their symbols.

Another important consideration is that *all* voltages and currents are expressed as '**root-mean-square**' values. But, as explained in an earlier chapter, we *don't* need to label them as such.

Purely-resistive circuit

So let's start with the simplest 'ideal' circuit: that is, a **purely-resistive** circuit (figure 3.1).

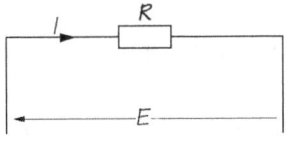

figure 3.1

In a **purely-resistive** circuit, the current $(\overline{I})$ and the potential difference $(\overline{E})$ across that resistance are described as being '**in phase**' with each other —i.e. the peak and zero points of their two separate waveforms correspond *exactly* throughout each complete cycle (figure 3.2). This is exactly how we would intuitively expect any circuit to behave but, as we shall learn, this is *not* the case with other types of circuit.

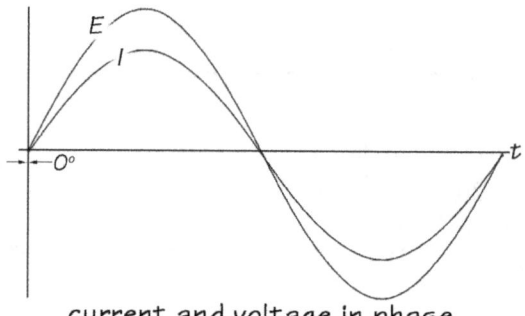

current and voltage in phase

figure 3.2

These waveforms can be represented by means of a '**phasor diagram**', with the current- and voltage-phasors each lying alongside each other. In this case, it's usual to draw them both along the horizontal real (positive) axis —i.e. horizontally, pointing to the right. In the following illustration, and in those that follow, the small curved arrow to the right is used simply to remind ourselves that, by common consent, phasors always 'rotate' in a counter-clockwise direction, but we have 'frozen' them in the positions shown so that we can analyse them.

Note that there are *two* ways of representing phasors which are in phase with each other: we can either *superimpose* one over the other (so that only the arrow-heads are distinguishable) or, as shown in figure 3.3, we can draw them with a very narrow gap between them. In this book, we use the latter method, as shown below.

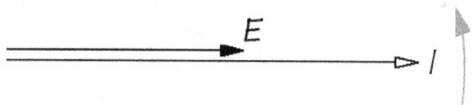

figure 3.3

The *length* of the voltage phasor $(\overline{E})$ represents the *r.m.s. value of the supply voltage*, and the *length* of the current phasor $(\overline{I})$ represents the *r.m.s. value of the current*. There is absolutely *no* relationship whatsoever between the scales of these two phasors, because they each represent *different* quantities (for example, if we were drawing them to scale, then the voltage phasor could be drawn to a scale of, say, 10 volts per millimetre, while the current phasor is drawn to a scale of, say, 2 amperes per millimetre). What *is* important, however, is the **angle** (or, in this particular case, the *lack* of any angle) between the two phasors, which indicates whether or not the two quantities are 'in phase' with each other.

As is the case for *any* electrical circuit, the ratio of voltage to current represents the *opposition* to current. In a **purely resistive circuit**, this 'opposition' is, of course, the **resistance** (symbol: R) of the circuit, measured in ohms:

$$R = \frac{\overline{E}}{\overline{I}}$$

It's important to remember that this equation tells us what the resistance *happens to be for that particular ratio of voltage to current* —the *actual* resistance itself being determined by the circuit's physical factors (length, cross-sectional area, and resistivity).

Note that we do *not* place a 'bar' over the resistance symbol, because resistance isn't a phasor quantity.

Purely-inductive circuit

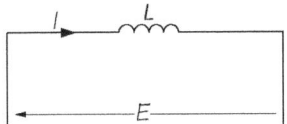

figure 3.4

Whenever the current through an **inductor** (figure 3.4) varies, a voltage is induced into that inductor through the process of *self-induction*. By Lenz's Law, the *direction* of this induced voltage *(u)* always acts to *oppose any change in current*. And, in accordance with Faraday's Law, this induced voltage is *directly proportional to the rate of change of current*, as expressed below:

$$u \propto -\frac{\Delta i}{\Delta t}$$

...where the Greek letter, 'delta' (Δ) simply means *'change in'*, so the expression $\Delta i / \Delta t$ means *'rate of change of current'*.

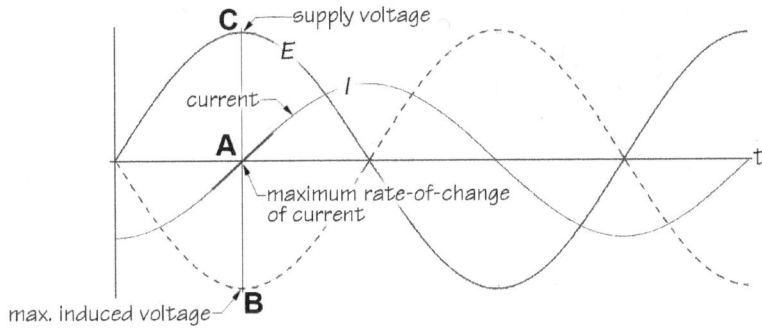

figure 3.5

In this case, the *greatest* rate of change in current occurs when the current waveform is at its *steepest*. As can be seen in figure 3.5, this occurs whenever the current waveform passes through the zero axis. So, at point **A**, for example, the current is *increasing* at its greatest rate of change, so the maximum self-induced voltage (point **B**) will occur at the same time but, in accordance with Lenz's Law, must act in the *negative* sense (i.e. opposing the increase in current). This induced voltage waveform is shown as a broken line in figure 3.5.

By **Kirchhoff's Voltage Law**, the induced voltage must be equal to, but *opposite*, the supply voltage (point **C**). The supply-voltage waveform is shown as a solid line. So, the current is clearly one-quarter of a wavelength *behind* the supply voltage $(\overline{E})$. We say, therefore, that *'the current is **lagging** the supply voltage by an angle of 90°'*.

So, in a **purely-inductive** circuit, then, the current $(\overline{I})$ *always* **lags** the supply voltage $(\overline{E})$ by 90° (and it is equally true to say that the *voltage always leads the current by 90°*). If we now redraw the waveform, ignoring the self-induced voltage, it will look like figure 3.6:

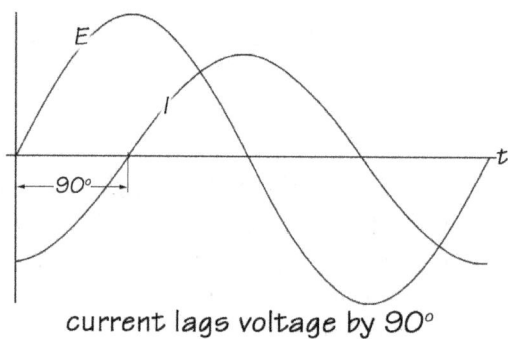

current lags voltage by 90°

figure 3.6

Again, the waveforms can be represented by means of a **phasor diagram**, with the current- and voltage-phasors *each lying at right-angles to each other* (figure 3.7). By convention, phasors are considered to 'rotate' counter-clockwise, so the voltage phasor $(\overline{E})$ is drawn 90° counter-clockwise from the current phasor $(\overline{I})$, as shown below. It would have been equally correct to have placed the voltage phasor horizontally, with the current phasor drawn 90° clockwise but, for consistency throughout this chapter, we'll draw the current phasor horizontally.

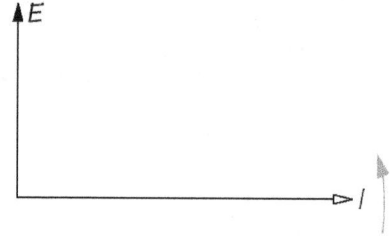

figure 3.7

As before, the *length* of the voltage phasor represents the *r.m.s. value of voltage*, and the *length* of the current phasor represents the *r.m.s. value of the current*. Again, there is absolutely *no* relationship between the lengths of the two phasors, as they are drawn to difference scales, with one representing voltage and the other current. What *is* important, however, is that *the voltage phasor is drawn 90° counter-clockwise relative to the current phasor*.

Once again, the ratio of voltage to current represents the *opposition* to current. In a **purely inductive circuit**, of course, *there is no resistance*, so it's necessary to call this 'opposition' by some other name. And, in fact, we call this 'opposition' to current **'inductive reactance'** (symbol: X_L) measured in ohms, with the term, 'reactance', meaning *'reacting against'* the passage of current.

$$X_L = \frac{\overline{E}}{\overline{I}}$$

Once again, it's important to understand that the ratio of voltage to current tells us what the inductive reactance happens to be *for that particular ratio*. The **inductive reactance** itself is determined by the *inductance* of the load, and the *frequency* of the supply, as shown in the following equation:

$$X_L = 2\pi f L$$

where: X_L = inductive reactance (ohms)
f = supply frequency (hertz)
L = inductance (henrys)

Unfortunately, the derivation of this equation is beyond the scope of this book. ***So it is necessary to commit this equation to memory***. But, don't worry, it's only the first of just *two* equations we will need to memorise!

Purely-capacitive circuit

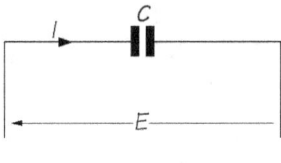

figure 3.8

For **capacitance** (figure 3.8), the instantaneous current *(i)* is *directly proportional to the rate of change of voltage*, as expressed below:

$$i \propto \frac{\Delta e}{\Delta t}$$

…where $\Delta e/\Delta t$ simply means 'rate of change of voltage.

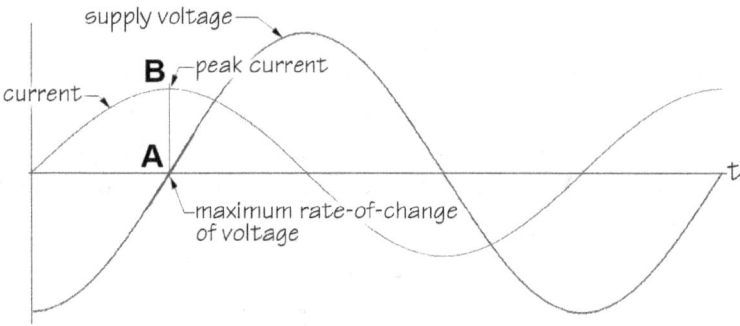

figure 3.9

In figure 3.9, the greatest rate of change of voltage occurs when the supply voltage waveform passes through the zero axis and is at its *steepest* —for example, at point **A**. So, this is the point at which the maximum current occurs (point **B**). In this case, the current is clearly one-quarter of a waveform *ahead* of the supply voltage. So we can say that *'the current **leads** the supply voltage by 90°'*.

In a **purely-capacitive** circuit, then, the current $(\overline{I})$ is said to **lead** the voltage-drop $(\overline{E})$ by 90° (and it is equally true to say the *voltage lags the current by 90°*).

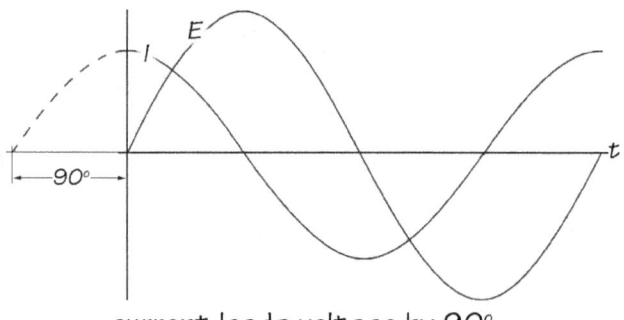

current leads voltage by 90°

figure 3.10

Again, the waveforms in figure 3.10 can be represented by means of a **phasor diagram**, with the current and voltage phasors each lying at right-angles to each other. The voltage phasor $(\overline{E})$ is drawn 90° *clockwise* relative to the current phasor $(\overline{I})$, as illustrated in figure 3.11. Again, it would be equally correct to draw the voltage phasor horizontally, with the current phasor 90° counter-clockwise but, for the sake of consistency, we have chosen to place the current phasor in the horizontal position,

figure 3.11

As with *all* phasor diagrams, the *length* of the voltage phasor represents the *r.m.s. value of voltage*, and the *length* of the current phasor represents the *r.m.s. value of the current*, but there is absolutely *no* relationship between the scales of the two phasors. What *is* important, however, is that *the voltage phasor is drawn 90° clockwise relative to the current phasor.*

As always, the ratio of voltage to current determines the *opposition* to current. In a **purely capacitive circuit**, there is no resistance or inductive reactance, so we need yet another term to describe it. So we call the 'opposition' to current the '**capacitive reactance**' (symbol: X_C) of the circuit, measured in ohms:

$$X_C = \frac{\overline{E}}{\overline{I}}$$

The ratio of voltage to current tells us what the capacitive reactance happens to be *for that particular ratio*. The **capacitive reactance** itself, however. is inversely-proportional to the *capacitance* of the load, and the *frequency* of the supply, as specified in the following equation:

$$X_C = \frac{1}{2\pi f C}$$

where: X_C = capacitive reactance (ohms)
f = supply frequency (hertz)
C = capacitance (farads)

Once again, the derivation of this equation is beyond the scope of this text. So, again, *it is necessary to commit this equation to memory*.

However, this equation, together with that for inductive reactance, *are the only two equations you will be asked to commit to memory!* From now on, *all* other equations can be derived from phasor diagrams!

The mnemonics 'CIVIL' or 'ELI the ICEman'

It is ***absolutely essential*** that we remember the phase-relationships between voltages and currents for **purely-resistive, purely-inductive, and purely-capacitive** circuits. Otherwise, *we will **not** be able to construct phasor diagrams!*

And if we cannot construct phasor diagrams, then we will <u>never</u> understand the behaviour of alternating current!

So, to help us, we should remember the mnemonic '***CIVIL***', in which '***C***' stands for 'capacitive circuit', and '***L***' stands for 'inductive circuit' (see figure 3.12).

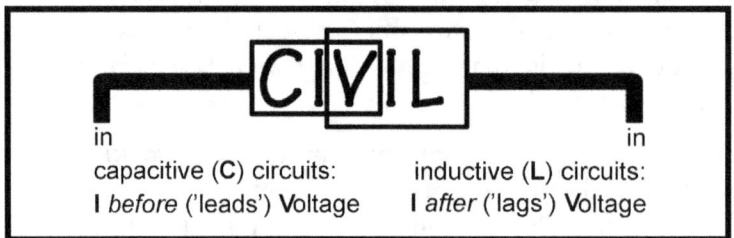

figure 3.12

An alternative mnemonic that we might prefer to remember is '*ELI the ICEman*', where '***ELI***' indicates that, in an inductive *(L)* circuit, voltage *(E)* is before (leads) current *(I)*; and where '***ICE***' indicates that, in a capacitive *(C)* circuit, current *(I)* is before (leads) voltage *(E)*.

Now that we have learned how these three 'ideal' (theoretical) circuits behave, let's now turn our attention to 'real' a.c. circuits.

So, what do we mean by 'real' a.c. circuits?

Well, *all* 'real' a.c. circuits exhibit *combinations* of resistance, inductance, and/or capacitance.

Since *all* real circuits exhibit resistance, we will look at **series resistive-inductive (series *R-L*)** circuits, then at **series resistive-capacitive (series *R-C*)** circuits and, finally, at **series resistive-inductive-capacitive (series *R-L-C*)** circuits.

Again, it's worth reminding ourselves that the circuit symbols used throughout this book represent the *quantities* resist<u>ance</u>, induct<u>ance</u>, and capacit<u>ance</u> —*not* resist<u>ors</u>, induct<u>ors</u>, and capacit<u>ors</u>.

Series R-L circuits

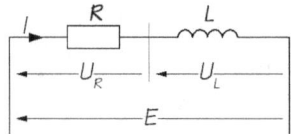

figure 3.13

The circuit diagram shown in figure 3.13, represents a **series resistive-inductive** circuit. It's *not* intended to represent a *resistor* in series with an *inductor*, but a load (such as a coil) which exhibits both *resistance* and *inductance*. We have, if you like, 'separated out' or 'split apart' the quantities, 'resistance' and 'inductance', of a coil or other inductive component, so that each can be considered separately.

We have already learnt that, in a purely resistive circuit, *the current and voltage are in phase* with each other and, in a purely inductive circuit, *the current **lags** the voltage by 90°*. So, what happens in a series *R-L* circuit? Well, our instinct should tell us that *the current is likely to lag the voltage by some angle between 0° and 90°* — and this, indeed, is the case. This angle we call the **'phase-angle'** (symbol: ϕ, pronounced *'phi'*) of the circuit.

The general definition of **phase angle** is *'the angle by which the **current** leads or lags the supply-voltage'*. Note, we *always* measure phase angles in terms of what the load *current* is doing, relative to the supply voltage, *never the other way around*.

So, for an *R-L* circuit, because the current *always* lags the supply-voltage, the phase-angle is *always* described as **lagging**. Whenever a phase-angle is quoted, *it is usual to specify whether it's leading or lagging*.

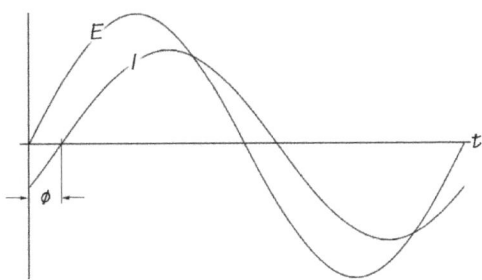

current lags voltage by angle ϕ

figure 3.14

Now, whenever a current $(\overline{I})$ flows through a **series *R-L* circuit**, a voltage-drop, $\overline{U}_R$, will appear across the resistive component of the circuit, and a voltage-drop, $\overline{U}_L$, will appear across the inductive component of the circuit —as shown in the schematic diagram in figure 3.13.

Voltage Phasor Diagram

Drawing the Phasor Diagram
Step 1:
In a series *R-L* circuit, the **current** is common to *both* the resistive *and* the inductive components and, so, current is *always* chosen as the **reference phasor**.

> **Important!** In **series** circuits, the same current is common to each component within the circuit, so **current** is *always* chosen as the reference phasor.

The reference phasor (figure 3.15) is *always drawn first, and it's **always** drawn along the real (horizontal) positive axis*. It's also drawn fairly long in order to distinguish it from the other phasors we are about to draw. In the following diagrams, we will further distinguish it by using an outline, rather than solid, arrow head (although this is not absolutely necessary).

figure 3.15

Step 2:

Since, as we have learnt, the voltage-drop, $\overline{U}_R$, across a resistive component is *in phase with the current,* it is also drawn along the horizontal positive axis (figure 3.16). Now, some textbooks show the phasor $\overline{U}_R$ *superimposed* over the reference phasor; others show it drawn *very close and parallel* with the reference phasor —the method preferred in this book:

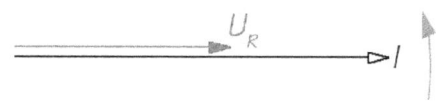

figure 3.16

Step 3:

As we have learnt, the voltage-drop, $\overline{U}_L$, across the inductive component *leads* the current by 90° (remember CI**VIL** or **ELI**), so is drawn 90° counter-clockwise ('leading') from the reference phasor —as shown in figure 3.17:

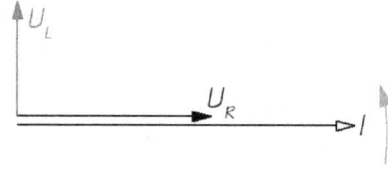

figure 3.17

Step 4:

We know, from **Kirchhoff's Voltage Law**, that in a series circuit, the supply voltage is the *sum of the individual voltage-drops*. However, because, in this case, the two voltage-drops, $\overline{U}_R$ and $\overline{U}_L$, lie at right-angles to each other, we have to add them *vectorially* (figure 3.18).

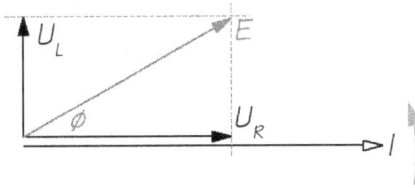

figure 3.18

From the completed phasor diagram (figure 3.18) we can see that the supply voltage, $\overline{E}$, is the **phasor-sum** (or **vectorial sum**) of $\overline{U}_R$ and $\overline{U}_L$, which can easily be determined using Pythagoras's Theorem:

$$\overline{E} = \sqrt{\overline{U}_R^2 + \overline{U}_L^2}$$

It's completely unnecessary to commit this equation to memory, because it has been derived from the phasor diagram, using Pythagoras's Theorem. If you can draw the phasor diagram, and know Pythagoras's Theorem, *then you don't need to bother to remember this equation!*

Worked Example 1

The voltage-drop across the resistive component of a series *R-L* circuit is 30 V, and the voltage-drop across the inductive component is 40 V. What is the value of the supply voltage?

Solution

Always start by sketching the circuit diagram (figure 3.19), and inserting all the values given to you in the question:

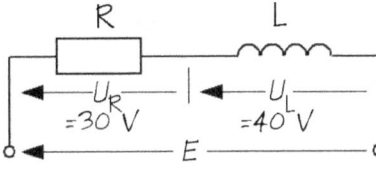

figure 3.19

Next, sketch the phasor diagram, following the steps described above. You *don't* have to draw the phasor diagram to scale (figure 3.20).

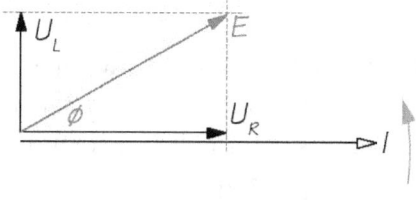

figure 3.20

Now, you can apply Pythagoras's Theorem to solve the problem:

$$\overline{E} = \sqrt{\overline{U}_R^2 + \overline{U}_L^2}$$
$$= \sqrt{30^2 + 40^2}$$
$$= \sqrt{2500} = 50 \text{ V (Answer)}$$

Impedance Diagram

The current flowing through a series *R-L* circuit will be opposed by *both* resistance (*R*) *and* the inductive reactance (X_L). The *combination* of these two different types of 'opposition' is called the **impedance** (symbol: **Z**) of the circuit, also measured in ohms. 'Impedance' is yet another word, meaning to 'oppose' or 'impede' the passage of current.

> Unlike voltages and currents, the quantities resistance, inductive reactance, and impedance are *not*, themselves, considered to be 'phasors'. However, the relationship between them is *vectorial* because they are the 'byproducts', if you like, of a 'voltage' phasor diagram.

However, we *cannot* simply add the resistance and inductive reactance —so how can we find the impedance? The answer is by means of an **impedance diagram.**

Drawing the Impedance Diagram
Step 1:

We start by drawing the circuit's phasor diagram, following the steps already explained (figure 3.21).

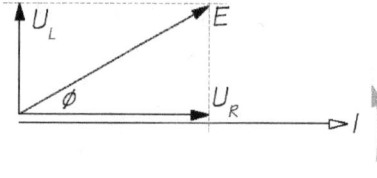

figure 3.21

Step 2:
Next, we *divide each of the voltage phasors by the reference phasor (**I**)* (figure 3.22).

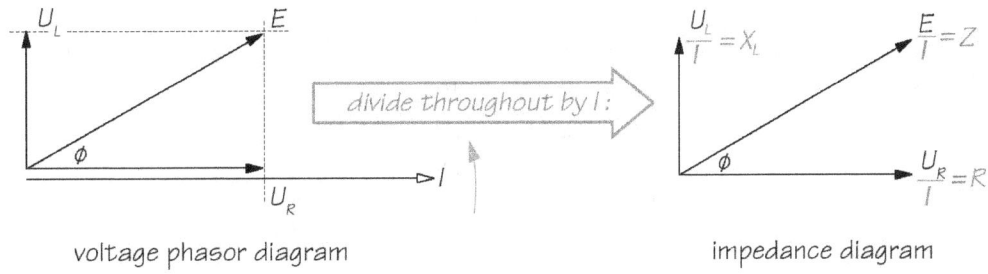

voltage phasor diagram   impedance diagram

figure 3.22

The resulting diagram is known as an '**impedance diagram**' (also called an '*impedance triangle*'), and is useful because it generates the following important equations:

$$R = \frac{\overline{U}_R}{\overline{I}} \qquad X_L = \frac{\overline{U}_L}{\overline{I}} \qquad Z = \frac{\overline{E}}{\overline{I}}$$

Again, *we don't have to commit these equations to memory*, because they are derived when we convert the voltage phasor diagram into an impedance diagram!

Also from the impedance diagram, we can see that the impedance is also the *vector sum of resistance and inductive reactance*, which can be calculated by simply applying Pythagoras's Theorem (figure 3.23).

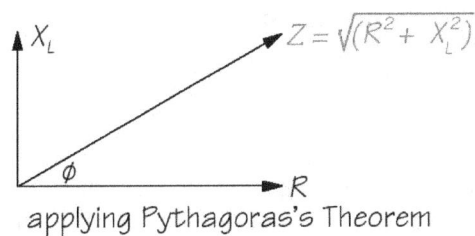

applying Pythagoras's Theorem

figure 3.23

$$Z = \sqrt{(R^2 + X_L^2)}$$

If required, we can also write similar equations for resistance and inductive reactance, by manipulating Pythagoras's Theorem. That is:

$$R = \sqrt{(Z^2 - X_L^2)} \qquad X_L = \sqrt{(Z^2 - R^2)}$$

Once again, we don't need to commit any of these equations to memory, *providing* we can draw a phasor diagram, convert it to an impedance diagram, and apply Pythagoras's Theorem!

We can also find the circuit's **phase-angle**, using basic trigonometry, utilising either the *sine*, *cosine*, or *tangent* ratios. In practice, for a reason we'll see later in this text, the best choice is always to use the *cosine* ratio:

$$\cos\phi = \frac{\text{adjacent}}{\text{hypotenuse}} = \frac{R}{Z}$$

Important!

Dividing a **voltage phasor** diagram by current produces an **impedance diagram** which generates each of the equations shown above. *So we don't have to learn any of these equations* —they can all be generated provided we learn how to draw the phasor and impedance diagrams!

Worked Example 2

An inductor, of resistance 5 Ω and inductance 0.02 H, is connected across a 230-V, 50 Hz, A.C. supply. Calculate each of the following:

    a.   inductive reactance

    b.   impedance

    c.   current

    d.   voltage-drop across the resistive component of the circuit

    e.   voltage-drop across the inductive component of the circuit

    f.   phase angle of the circuit

Solution

As always, the first step in solving any a.c. circuit problem is to sketch the circuit diagram, and label it with all values supplied in the problem (figure 3.24).

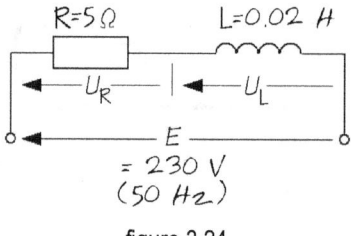

figure 3.24

The next step is to construct the voltage phasor diagram, following the steps described earlier (figure 3.5).

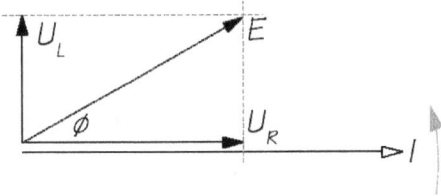

figure 3.25

As the problem relates to inductive reactance, impedance, etc., the next step is to convert the voltage phasor diagram into an **impedance diagram** by *dividing throughout by the reference quantity* —i.e. by the current. This generates all the equations that we need to solve the problem (figure 3.26).

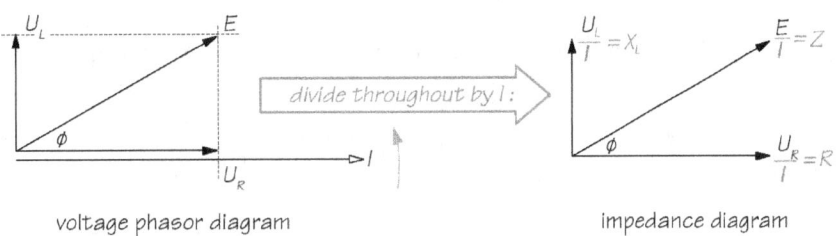

voltage phasor diagram   impedance diagram

figure 3.26

a. To find the **inductive-reactance** (X_L) of the circuit, we start by looking at the equations generated when we constructed the impedance diagram. There's only one equation for inductive reactance, $X_L = \bar{U}_L / I$. Unfortunately, we don't know the value of $\bar{U}_L$, so we *cannot* use this formula. What about applying Pythagoras's Theorem ($X_L = \sqrt{Z^2 - R^2}$)? Could we use this equation to find X_L? Unfortunately, no, because we don't know the value of Z. If we *can't* use any of the equations generated by the impedance diagram, then we *must fall back on the basic equation for inductive reactance*, as follows:

$$X_L = 2\pi f L = 2\pi \times 50 \times 0.02 = 6.28\ \Omega \text{ (Answer a.)}$$

b. To find the impedance, we *can* use an equation generated by the impedance diagram:

$$Z = \sqrt{R^2 + X_L^2}$$
$$= \sqrt{5^2 + 6.28^2}$$
$$= \sqrt{25 + 39.44}$$
$$= \sqrt{64.44} = 8.03\ \Omega \text{ (Answer b.)}$$

c. To find the current, we use the following equation that was generated by the impedance diagram:

$$\overline{I} = \frac{\overline{E}}{Z} = \frac{230}{8.03} = 28.64 \text{ A (Answer c.)}$$

d. Again, using the equation generated by the impedance diagram:

$$\overline{U}_R = \overline{I}R = 28.64 \times 5 = 143.20 \text{ V (Answer d.)}$$

e. Again, using the equation generated by the impedance diagram:

$$\overline{U}_L = \overline{I}X_L = 28.64 \times 6.28 = 179.86 \text{ V (Answer e.)}$$

f. Using the cosine function:

$$\angle\phi = \cos^{-1}\frac{R}{Z} = \cos^{-1}\frac{5}{8.03} = \cos^{-1}0.6227 = 51.48° \text{ lagging (Answer f.)}$$

(*'Lagging'*, because the supply current lags the supply voltage in an inductive circuit.)

> **Note**, the symbol, '**cos⁻¹**' is simply mathematicians' shorthand for the phrase, *'The angle whose cosine is...'*

Series R-C circuits

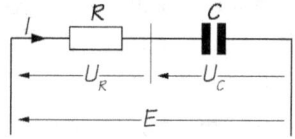

figure 3.27

Again, it's important to realise that the circuit, shown in figure 3.27, does not necessarily represent a *resistor* in series with a *capacitor*; rather, it simply represents the *quantities* resistance in series with capacitance. For example, it could represent the resistance and capacitance of a very long underground cable.

We know that in a purely resistive circuit, the current and voltage are in phase with each other; and, in a purely capacitive circuit, the current leads the voltage by 90°. So, what happens in a *series R-C* circuit? Well, clearly, this time we instinctively know that the *current must lead the voltage by some angle between 0° and 90°* —this angle is called the circuit's **phase-angle** (symbol: ϕ, pronounced 'phi'), as illustrated in figure 3.28.

Remember, the general definition of **phase angle** is *the angle by which the current leads or lags the supply-voltage*. So for *resistive-capacitive circuits*, because the current *always* leads the supply-voltage, the phase-angle is *always* described as **leading**.

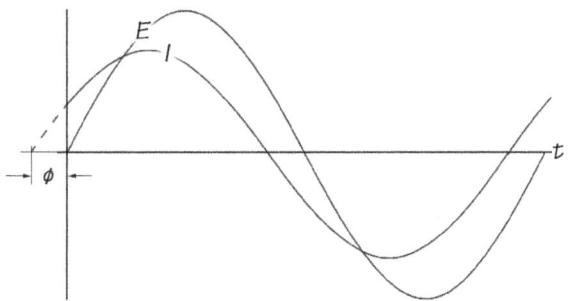

current leads voltage by angle ϕ

figure 3.28

When a current $(\overline{I})$ flows through a **series R-C circuit**, a voltage-drop $\overline{U}_R$ will appear across the resistive component of the circuit, and a voltage-drop $\overline{U}_C$ will appear across the capacitive component of the circuit.

Voltage Phasor Diagram

Drawing the Voltage Phasor Diagram

Step 1:

In a series circuit, the **current** is common to each component and, so, current is *always* chosen as the **reference phasor**. The reference phasor is *always drawn along the horizontal positive axis*, and it's also normally drawn fairly long in order to distinguish it from the others (figure 3.29).

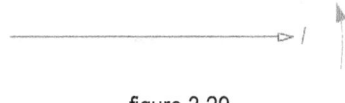

figure 3.29

Step 2:

The voltage-drop, $\overline{U}_R$, across the resistive component is *in phase with the current* and, so, is also drawn along the horizontal positive axis (figure 3.30).

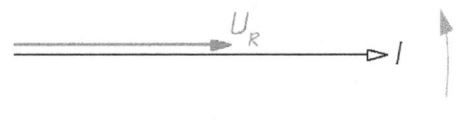

figure 3.30

Step 3:

The voltage-drop, $\overline{U}_C$, across the capacitive component *lags the current by 90°* (remember **CIV**IL or **ICE**man), so is drawn 90° clockwise from the reference phasor (figure 3.31).

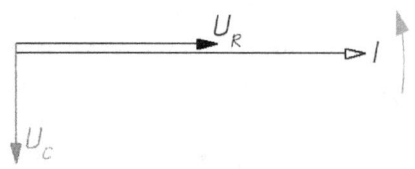

figure 3.31

Step 4:

We know, from Kirchhoff's Voltage Law, that in a series circuit, the total voltage-drop is the sum of the individual voltage-drops. Because, in this case, the two voltage-drops, $\overline{U}_R$ and $\overline{U}_C$, lie at right-angles to each other, we have to add them *vectorially*:

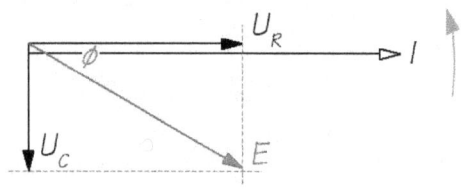

figure 3.32

From the completed phasor diagram shown in figure 3.32, we can see that the supply voltage, $\overline{E}$, is the **phasor-sum** (or **vectorial sum**) of $\overline{U}_R$ and $\overline{U}_C$, which can be found using Pythagoras's Theorem:

$$\overline{E}=\sqrt{\overline{U}_R^2+\overline{U}_C^2}$$

Worked Example 3

The voltage-drop across the resistive component of a series *R-C* circuit is 40 V, and the voltage-drop across the capacitive component is 30 V. What is the value of the total voltage-drop?

Solution

Always start by sketching the circuit diagram (figure 3.33), and inserting all the values given to you in the question:

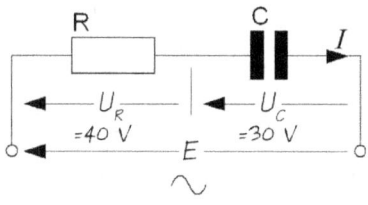

figure 3.33

Next, sketch the phasor diagram, following the steps described above. You *don't* have to draw the phasor diagram to scale (figure 3.34).

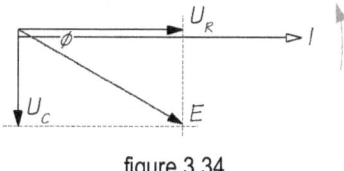

figure 3.34

Now, you can apply Kirchhoff's Voltage Law, and use Pythagoras's Theorem to solve the problem:

$$\overline{E} = \sqrt{\overline{U}_R^2 + \overline{U}_C^2}$$
$$= \sqrt{40^2 + 30^2}$$
$$= \sqrt{2500} = 50 \text{ V (Answer)}$$

Impedance Diagram

The current flowing through a series *R-C* circuit will be opposed by both its resistance (*R*) *and* by its capacitive reactance (X_C). The *combination* of these two types of 'opposition' is called the **impedance** (symbol: **Z**) of the circuit, and is measured in ohms. Again, we *cannot* simple add the resistance and capacitive reactance —so how can we find the impedance? Again, the answer is by means of an **impedance diagram**:

Drawing the Impedance Diagram
Step 1:

We start by drawing the circuit's phasor diagram, following the steps already explained (figure 3.35).

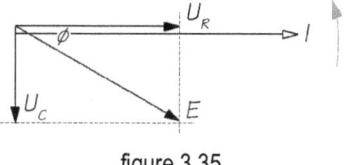

figure 3.35

Step 2:

Next, we *divide each of the voltage phasor by the reference phasor ($\overline{I}$)*:

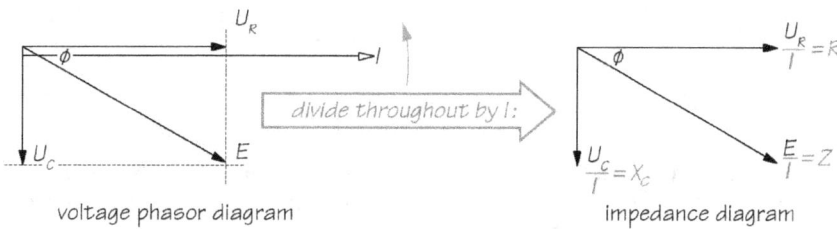

figure 3.36

The resulting diagram (figure 3.26) is an **impedance diagram** (or *'impedance triangle'*), and is useful because it generates the following equations:

$$R = \frac{\overline{U}_R}{\overline{I}} \qquad X_C = \frac{\overline{U}_C}{\overline{I}} \qquad Z = \frac{\overline{E}}{\overline{I}}$$

Also from the impedance diagram, we can also see that the impedance is also the *vectorial sum of resistance and capacitive reactance*, which can be calculated by applying Pythagoras's Theorem (figure 3.37).

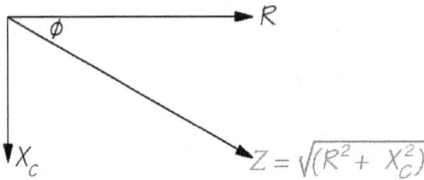

applying Pythagoras's Theorem

figure 3.37

$$Z = \sqrt{R^2 + X_C^2}$$

We can, if necessary, write similar equations for the resistance and capacitive reactance, by applying Pythagoras's Theorem. That is:

$$R = \sqrt{(Z^2 - X_C^2)} \qquad X_C = \sqrt{(Z^2 - R^2)}$$

We can also find the circuit's **phase-angle**, using basic trigonometry, utilising either the *sine*, *cosine*, or *tangent* ratios —but, as before, the best choice is to use the *cosine* ratio:

$$\cos \phi = \frac{\text{adjacent}}{\text{hypotenuse}} = \frac{R}{Z}$$

$$\angle \phi = \cos^{-1} \frac{R}{Z}$$

Important!

Dividing a **voltage phasor diagram** by current produces an **impedance diagram** which generates each of the equations shown above. *So we don't have to learn any of these equations —they can all be generated provided we learn how to draw the phasor and impedance diagrams!*

Worked Example 4

A capacitor, of resistance 40 Ω and capacitance 50 μF is connected across a 110-V, 50 Hz, A.C. supply. Calculate each of the following:

    a.   capacitive reactance
    b.   impedance
    c.   current
    d.   voltage-drop across the resistive component of the circuit
    e.   voltage-drop across the capacitive component of the circuit
    f.   phase angle of the circuit

Solution

The first step in solving any a.c. circuit problem is to sketch the circuit diagram, and label it with all values supplied in the problem (figure 3.38).

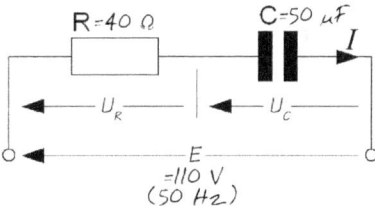

figure 3.38

The next step is to draw the voltage phasor diagram (figure 3.39) following the steps described earlier.

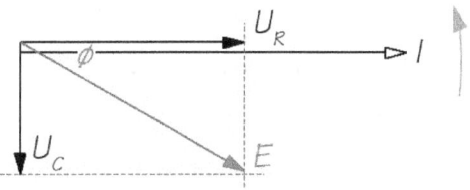

figure 3.39

As the problem relates to inductive reactance, impedance, etc., the next set is to convert the voltage phasor diagram into an impedance diagram, by dividing throughout by the reference quantity —i.e. current. This generates the equations that we need to solve the problem (figure 3.40).

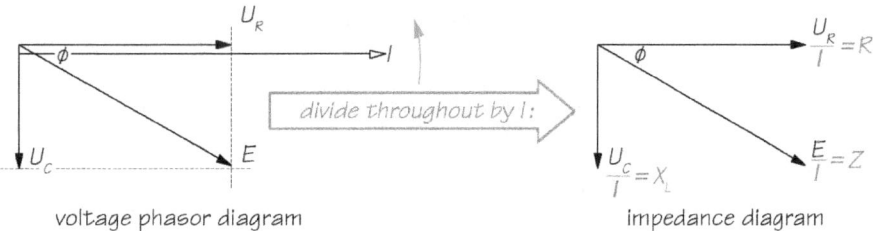

voltage phasor diagram   impedance diagram

figure 3.40

a. To find the **capacitive-reactance** (X_C) of the circuit, we start by looking at the equations generated when we constructed the impedance diagram. There's only one equation for capacitive reactance, $X_C = \bar{U}_C / \bar{I}$. Unfortunately, we don't know the value of $\bar{U}_C$ so we can't use this formula. What about applying Pythagoras's Theorem ($X_C = \sqrt{Z^2 - R^2}$)? Could we use this to find X_C? Unfortunately, no, because we don't know the value of Z. Clearly, then, we must fall back on the basic equation for capacitive reactance, as follows:

$$X_C = \frac{1}{2\pi f C} = \frac{1}{2\pi \times 50 \times (50 \times 10^{-6})} = 63.67 \, \Omega \text{ (Answer a.)}$$

b. To find the impedance, we can now use the equation generated by the impedance diagram:

$$Z = \sqrt{R^2 + X_L^2}$$
$$= \sqrt{40^2 + 63.67^2}$$
$$= \sqrt{1600 + 4054}$$
$$= \sqrt{5654} = 75.2 \, \Omega \text{ (Answer b.)}$$

c. To find the current, we use the following equation generated by the impedance diagram:

$$\bar{I} = \frac{\bar{E}}{Z} = \frac{110}{75.2} = 1.46 \text{ A (Answer c.)}$$

d. Again, using the equation generated by the impedance diagram:

$$\bar{U}_R = \bar{I} R = 1.46 \times 40 = 58.53 \text{ V (Answer d.)}$$

e. Again, using the equation generated by the impedance diagram:

$$\bar{U}_C = \bar{I} X_C = 1.46 \times 63.67 = 93.14 \text{ V (Answer e.)}$$

f. Using the cosine function:

$$\angle \phi = \cos^{-1} \frac{R}{Z} = \cos^{-1} \frac{40}{75.2} = \cos^{-1} 0.5319 = 57.87° \text{ leading (Answer f.)}$$

(*'Leading'*, because the supply current leads the supply voltage in a capacitive circuit)

Series R-L-C circuits

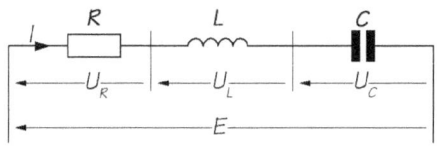

figure 3.41

So far, in this chapter, we have learnt that:

- in a **series R-L circuit**, *the current **lags** the supply voltage by some angle ϕ* and,

- in a **series R-C circuit**, *the current **leads** the supply voltage by some angle ϕ.*

So, what happens in a *series R-L-C* circuit? Well, clearly the *current could either lag or lead the voltage* —depending on the values of the circuit's inductive reactance and its capacitive reactance! Whatever value this angle happens to be, it will be the circuit's **phase-angle** (symbol: ϕ, pronounced 'phi').

When a current $(\overline{I})$ flows through a series R-L-C circuit, a voltage-drop $\overline{U}_R$ will appear across the resistive component of the circuit, a voltage-drop $\overline{U}_L$ will appear across the inductive component, and a voltage-drop $\overline{U}_C$ will appear across the capacitive component of the circuit.

Voltage Phasor Diagram

Drawing the Phasor Diagram
Step 1:
In a series circuit, the **current** is common to each component and, so, current is again chosen as the **reference phasor**. The reference phasor is *always drawn along the horizontal positive axis* and, as before, it's also normally drawn fairly long in order to distinguish it from the others (figure 3.42).

figure 3.42

Step 2:
The voltage-drop, $\overline{U}_R$, across the resistive component is *in phase with the current* and, so, is also drawn along the horizontal positive axis (figure 3.43).

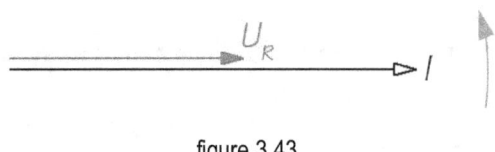

figure 3.43

Step 3:

The voltage-drop, $\bar{U}_L$, across the inductive component *leads the current by 90°* (remember C**IVIL** or **ELI**), so is drawn 90° counter-clockwise from the reference phasor (figure 3.44).

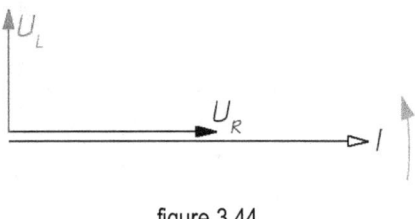

figure 3.44

Step 4:

The voltage-drop, $\bar{U}_C$, across the capacitive component *lags the current by 90°* (remember **CIV**IL or **ICE**man), so is drawn 90° clockwise from the reference phasor. In this case, we should *always* make the $\bar{U}_C$ phasor smaller than the $\bar{U}_L$ phasor, or *vice versa*. What we should *not* do, is to make them the same length as each other. We will see why not, in a moment.

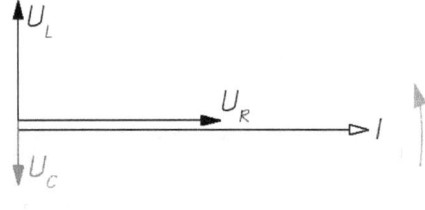

figure 3.45

Step 5:

From Kirchhoff's Voltage Law, the total voltage-drop in a series circuit, is the sum of the individual voltage-drops, and we have to add them *vectorially*.

It's a little more difficult to add *three* phasors. As $\bar{U}_L$ and $\bar{U}_C$ lie in *opposite* directions, the simplest thing to do is to start by subtracting them and, *then*, add the difference to phasor $\bar{U}_R$ (figure 3.46).

The snag is, of course, that we might not know, for a particular circuit, whether $\bar{U}_L$ is bigger than $\bar{U}_C$, or *vice-versa*! Fortunately, *it doesn't matter!* The purpose of the phasor diagram is simply to *generate equations*, not necessarily to accurately represent the actual conditions in the circuit to scale! And the phasor diagram will *always* generate the correct equations whether $\overline{U}_L$ is actually bigger than $\overline{U}_C$, or vice-versa!

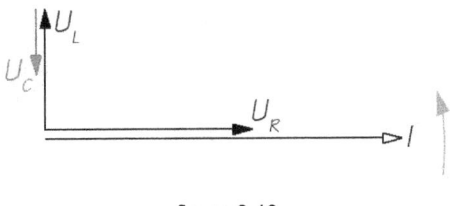

figure 3.46

So, the simplest solution is to get into the habit of ***always* drawing $\overline{U}_L$ longer than $\overline{U}_C$** (figure 3.46) —or the other way around, if you prefer! But, for a reason that will be revealed later, whatever you do, **never *ever* draw them the same length!!**

Figure 4.47 shows what the finished phasor diagram will look like:

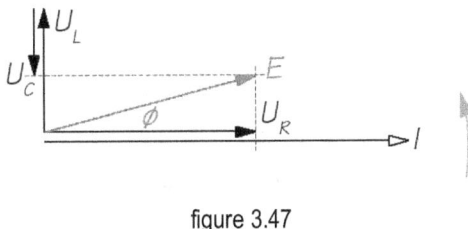

figure 3.47

From the completed phasor diagram, above, we can see that $\overline{E}$ is the **phasor-sum** (or **vector sum**) of $\overline{U}_R$, $\overline{U}_L$, and $\overline{U}_C$, which can be found using Pythagoras's Theorem:

$$\overline{E} = \sqrt{\overline{U}_R^2 + (\overline{U}_L - \overline{U}_C)^2}$$

If, when we draw the phasor diagram, we make $\overline{U}_L$ bigger than $\overline{U}_C$ when, really, it's the other way around, *it doesn't really matter*. When we subtract the two, we'll end-up with a negative quantity inside the brackets which, when squared, will result in a positive value. So, we'll *always* get the correct answer, even if we sketched the phasor diagram wrong!

Worked Example 5

In a series *R-L-C* circuit, the voltage-drop across the resistive component is 4 V, the voltage-drop across the inductive component is 10 V, and the voltage-drop across the capacitive component is 7 V. What is the value of the total voltage-drop?

Solution

Always start by sketching the circuit diagram, and inserting all the values given to you in the question (figure 3.48).

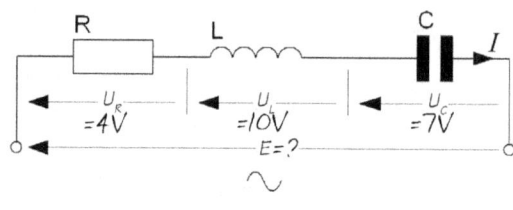

figure 3.48

Next, sketch the phasor diagram, following the steps described above. You *don't* have to draw the phasor diagram to scale (figure 3.49) as its only purpose is to generate equations, *not* to accurately represent the circuit.

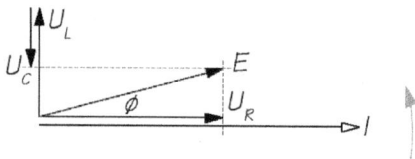

figure 3.49

Now, you can apply Kirchhoff's Voltage Law, and use Pythagoras's Theorem to solve the problem:

$$\overline{E} = \sqrt{\overline{U}_R^2 + (\overline{U}_L - \overline{U}_C)^2}$$
$$= \sqrt{4^2 + (10-7)^2}$$
$$= \sqrt{4^2 + 3^2} = \sqrt{25} = 5 \text{ V (Answer)}$$

Impedance Diagram

The current flowing through a series *R-L-C* circuit will be opposed by resistance (*R*) *and* by inductive reactance (X_L), *and* by capacitive reactance (X_C). The combination of these three different 'oppositions' is called the **'impedance'** (symbol: **Z**) of the circuit, measured in ohms. But, again, we *cannot* simple add the resistance, inductive reactance, and capacitive reactance —so how can we find the impedance? Again, the answer is by means of an **impedance diagram**:

Drawing the Impedance Diagram
Step 1:

We start by drawing the circuit's phasor diagram, following the steps already explained (figure 3.50).

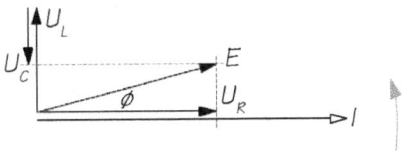

figure 3.50

Step 2:

Next, we *divide each of the individual voltage phasors by the reference phasor* $(\overline{I})$ (figure 3.51).

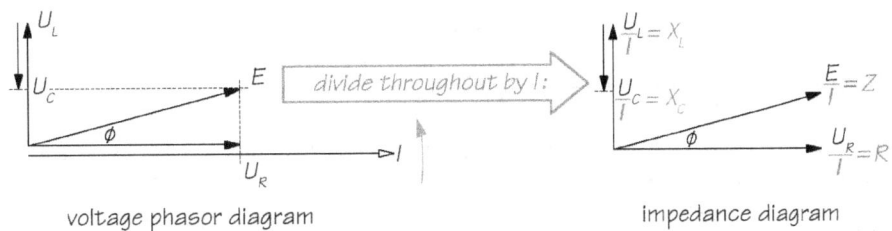

figure 3.51

The resulting diagram is an **impedance diagram** (or *'impedance triangle'*), and is useful because it generates the following equations:

$$R = \frac{\overline{U}_R}{\overline{I}}$$

$$X_L = \frac{\overline{U}_L}{\overline{I}}$$

$$X_C = \frac{\overline{U}_C}{\overline{I}}$$

$$Z = \frac{\overline{E}}{\overline{I}}$$

Also from the impedence diagram, we can see that the impedance is also the *vectorial sum of resistance, inductive reactance, and capacitive reactance*, which can be calculated by applying Pythagoras's Theorem (figure 3.52).

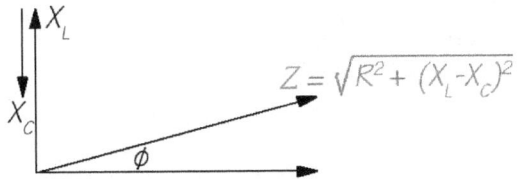

applying Pythagoras's Theorem

figure 3.52

$$Z = \sqrt{R^2 + (X_L - X_C)^2}$$

Once again, we can also find the circuit's **phase-angle**, using basic trigonometry:

$$\cos \phi = \frac{\text{adjacent}}{\text{hypotenuse}} = \frac{R}{Z}$$

$$\angle \phi = \cos^{-1} \frac{R}{Z}$$

Important!
Dividing a **voltage phasor diagram** by current produces an **impedance diagram** which generates each of the equations shown above. So you don't have to learn *any* of these equations —they can all be generated *provided you learn how to draw the phasor and impedance diagrams!*

Worked Example 6

A circuit of resistance of 1.5 Ω, inductance of 16 mH, and capacitance 500 μF, is connected across a 230-V, 50 Hz, A.C. supply. Calculate each of the following:

- a. inductive reactance
- b. capacitive reactance
- c. impedance
- d. current
- e. voltage-drop across the resistive component of the circuit
- f. voltage-drop across the inductive component of the circuit
- g. voltage-drop across the capacitive component of the circuit
- h. phase angle of the circuit

Solution

The first set in solving *any* A.C. circuit problem is to sketch the circuit diagram, and label it with all values supplied in the problem (figure 3.53).

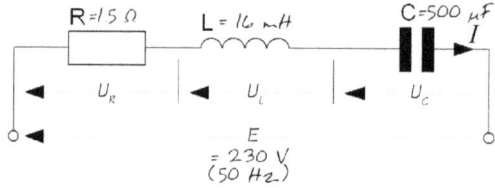

figure 3.53

The next step is to draw the voltage phasor diagram, following the steps described earlier (figure 3.54).

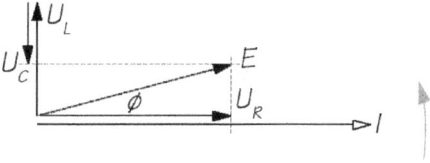

figure 3.54

As the problem relates to inductive reactance, capacitive reactance, impedance, etc., the next set is to convert the voltage phasor diagram into an *impedance diagram* (figure 3.55)., by dividing throughout by the reference quantity —i.e. current. This generates the equations that we need to solve the problem:

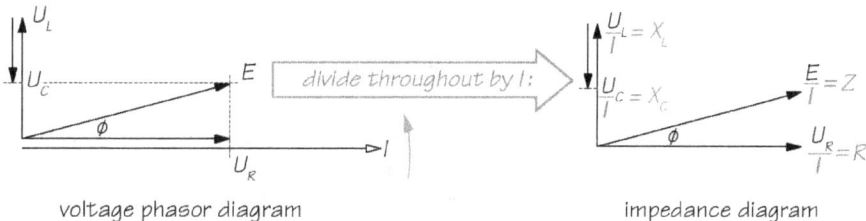

voltage phasor diagram   impedance diagram

figure 3.55

a.   To find the **inductive-reactance** (X_L) of the circuit, we start by looking at the equations generated when we constructed the impedance diagram. There's only one equation for inductive reactance, $X_L = \overline{U}_L / \overline{I}$. Unfortunately, we don't know the value of $\overline{U}_L$ so we can't use this formula. What about applying Pythagoras's Theorem? Could we use this to find X_L? Unfortunately, no, because we don't know the value of Z. Clearly, then, we must fall back on the basic equation for inductive reactance, as follows:

$$X_L = 2\pi f L = 2\pi \times 50 \times 16 \times 10^{-3} = 5.03 \, \Omega \text{ (Answer a.)}$$

b.   To find the **capacitive-reactance** (X_C) of the circuit, we start by looking at the equations generated when we constructed the impedance diagram. There's only one equation for capacitive reactance, $X_C = \overline{U}_C / \overline{I}$. Unfortunately, we don't know the value of $\overline{U}_C$ so we can't use this formula. What about applying Pythagoras's Theorem? Could we use this to find X_C? Unfortunately, no, because we don't know the value of Z. Clearly, then, we must fall back on the basic equation for capacitive reactance, as follows:

$$X_C = \frac{1}{2\pi f C} = \frac{1}{2\pi \times 50 \times (500 \times 10^{-6})} = 6.37 \, \Omega \text{ (Answer a.)}$$

c. To find the impedance, we can now use the equation generated by the impedance diagram:

$$Z = \sqrt{R^2 + (X_L - X_C)^2}$$
$$= \sqrt{1.5^2 + (5.03 - 6.37)^2}$$
$$= \sqrt{1.5^2 + (-1.34)^2}$$
$$= \sqrt{2.25 + 1.8}$$
$$= \sqrt{4.05} = 2.01 \, \Omega \text{ (Answer c.)}$$

Note! Despite X_C being larger than X_L, the above equation still delivers the correct answer, as the negative sign inside the bracket becomes positive when the bracket is squared!

d. To find the current, we use the following equation generated by the impedance diagram:

$$\overline{I} = \frac{\overline{E}}{Z} = \frac{230}{2.01} = 114.43 \text{ A (Answer d.)}$$

e. Again, using the equation generated by the impedance diagram:

$$\overline{U}_R = \overline{I}R = 114.43 \times 1.5 = 171.65 \text{ V (Answer e.)}$$

f. Again, using the equation generated by the impedance diagram:

$$\overline{U}_L = \overline{I}X_L = 114.43 \times 5.03 = 575.38 \text{ V (Answer f.)}$$

g. Again, using the equation generated by the impedance diagram:

$$\overline{U}_C = \overline{I}X_C = 114.43 \times 6.37 = 728.92 \text{ V (Answer g.)}$$

h. Using the cosine function:

$$\angle \phi = \cos^{-1}\frac{R}{Z} = \cos^{-1}\frac{1.5}{2.01} = \cos^{-1} 0.7462 = 41.74° \text{ leading (Answer h.)}$$

('Leading' because, *despite* how we have actually drawn the phasor diagram, $\overline{U}_C$ is, in this case, *larger* than $\overline{U}_L$, so the supply current must *lead* the supply voltage.)

Series resonance

From the following equations:

$$X_L = 2\pi f L \quad \text{and} \quad X_C = \frac{1}{2\pi f C}$$

...we know that the inductive reactance of a circuit is *directly*-proportional to the supply frequency, whereas the capacitive reactance is *inversely*-proportional to the supply frequency. So, if we had a variable-frequencey supply source and were to gradually *increase* the supply frequency to an *R-L-C* circuit, the inductive reactance would gradually *increase*, while the capacitive reactance woulf gradually *decrease*, following the curves shown in figure 3.56.

Eventually, a cross-over point will be reached, when the values of the inductive reactance and capacitive reactance will equal each other —as illustrated in figure 3.56.

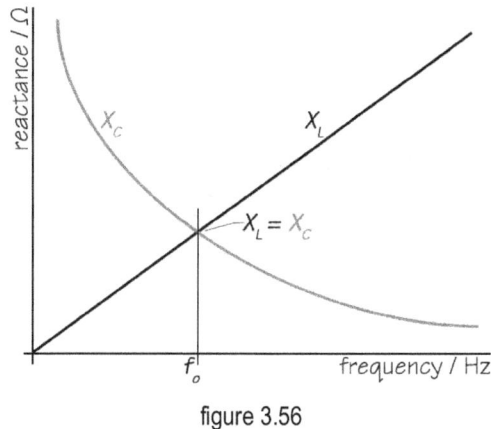

figure 3.56

The frequency at which this occurs is called the '**resonant frequency**' *(f_o)* of the circuit, and the circuit is said to exhibit '**series resonance**'.

'**Series resonance**', then, is a *unique condition* which *only* occurs whenever a series *R-L-C* circuit's inductive reactance is *exactly* equal to its capacitive reactance. That is, resonance occurs whenever:

$$X_L = X_C$$

As we shall see, at resonance, *rather strange things start to happen to a circuit!*

Any R-L-C circuit can be made to resonate at its unique resonant frequency *(f_o)*, and determining what that frequency is for any circuit is quite straightforward. Starting with the basic condition for resonance:

$$X_L = X_C$$

...and expanding:

$$2\pi f_o L = \frac{1}{2\pi f_o C}$$

$$f_o^2 = \frac{1}{(2\pi)^2 LC}$$

$$f_o = \frac{1}{2\pi \sqrt{LC}}$$

Worked Example 7

What is the resonant frequency for a circuit having an inductance of 16 mH and a capacitance of 100 µF?

Solution

$$f_o = \frac{1}{2\pi\sqrt{LC}} = \frac{1}{2\pi\sqrt{(16\times 10^{-3})\times(100\times 10^{-6})}}$$

$$= \frac{1}{2\pi\sqrt{1600\times 10^{-9}}} = \frac{1}{2\pi\times(1.26\times 10^{-3})} = 126 \text{ Hz (Answer)}$$

Impedance Diagram for Series Resonance

If we were to draw an **impedance diagram** for a series R-L-C circuit at resonance, it would look like figure 3.57.

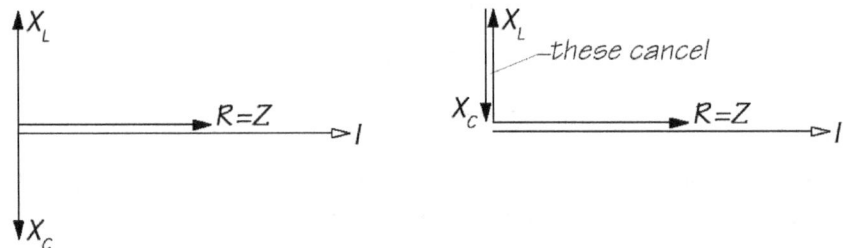

figure 3.57

As we can see, the inductive reactance and the capacitive reactance are *identical*, but *act in opposite directions*. So, their vectorial-sum must be zero.

This leaves **resistance** as the circuit's *only opposition to the flow of current*, from which we can say:

> At **resonance**, a circuit's impedance is equal to its resistance.

So if, at resonance, the vector-sum of a circuit's inductive reactance and capacitive reactance is zero, leaving only its resistance to oppose current, then it follows that the circuit's current will achieve its maximum value at resonance.

> A circuit's current reaches its maximum value when resonance occurs.

If we were to conduct a simple experiment, by adjusting the supply frequency of a series R-L-C circuit, until resonance occurs, while measuring its current, the resulting graph would look something like figure 3.58.

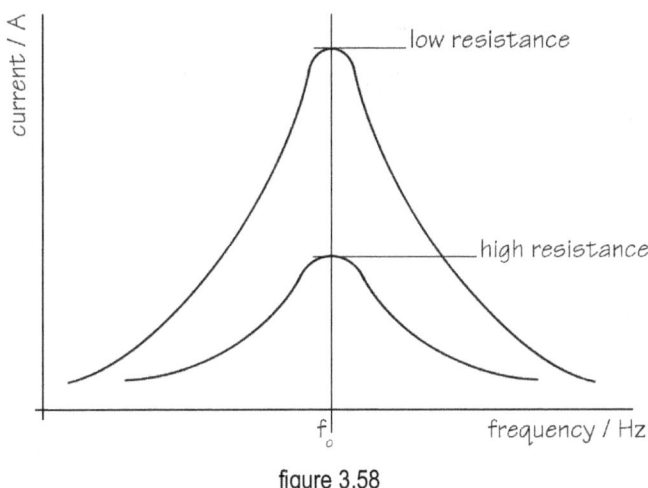

figure 3.58

As the frequency approaches the resonant frequency, the reactance (i.e. the combined inductive and capacitive reactance) falls towards zero ohms, and the corresponding current increases. The value which the current reaches, at resonance, is limited by the resistance of the circuit. If the resistance is *low*, then the resulting current will be *high*; if the resistance is *high*, then the resulting current will be *low*. As the applied frequency passes beyond the resonant frequency, the reactance starts to increase again, and the current starts to fall again.

Now, let's look at the **voltage phasor diagram** for an *R-L-C* circuit at resonance, shown in figure 3.59.

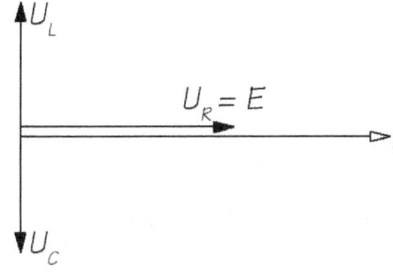

figure 3.59

As the voltage drops, $\overline{U}_L$ and $\overline{U}_C$, are equal and opposite, then the phasor sum of $\overline{U}_L$, $\overline{U}_C$, and $\overline{U}_R$ will simply be $\overline{U}_R$. In other words, $\overline{U}_R = \overline{E}$ or, to put in another way: the *entire supply voltage will appear across the resistive component of the circuit*.

> You will recall that when we learnt how to draw general voltage phasor diagrams for *R-L-C* circuits, we were warned *always* to make the $\overline{U}_L$ phasor larger than the $\overline{U}_C$ phasor *(or vice-versa)*. Well, now you know why! Making the two phasors the same lengths results in resonance, which is a *special* condition, not the normal condition!

It's very important to understand that we are *not* saying that the voltages $\overline{U}_L$ and $\overline{U}_C$ don't exist. They most certainly *do* exist! What we *are* saying is that, if we tried to measure the voltage drop across *both* the inductive *and* capacitive components, then a voltmeter would read zero because the two voltage drops are in *antiphase* —that is, they are equal in magnitude, but 180° out of phase with each other, as illustrated in figure 3.60.

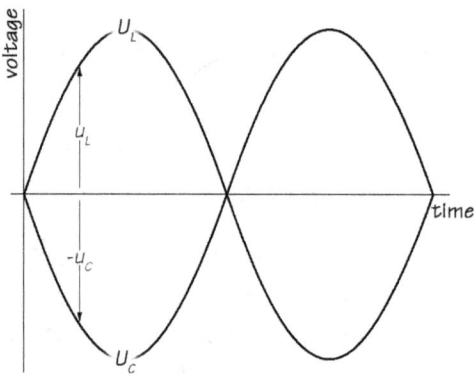

figure 3.60

So, at any instant, the sum of the instantaneous voltages, $u_L + (-u_C)$, must be zero. Or, in general:

$$\overline{U}_L + (-\overline{U}_C) = 0$$

This can be summarised as shown in figure 3.61 and table 3.1.

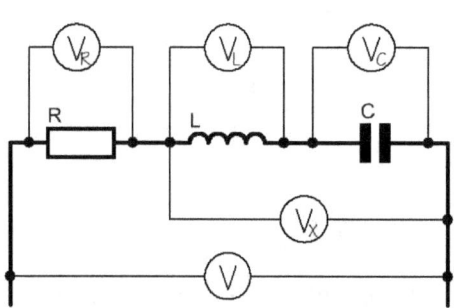

figure 3.61

Table 3.1

voltmeter V	measures the potential difference across the entire circuit.
voltmeter V_R	equals the reading of voltmeter **V**.
voltmeter V_L	measures the voltage drop across the inductive component.
voltmeter V_C	measures the voltage drop across the capacitive component.
voltmeter V_X	measures zero, because: $[V_L + (-V_C) = 0]$.

The voltage phasor diagram also tells us another important thing about resonance. At resonance, *the circuit's phase angle is zero.*

> At resonance the circuit's phase angle is zero.

The final 'strange' thing which occurs, or rather, which *could* occur, at resonance is revealed in the following worked example.

Worked Example 8

A circuit has a resistance of 10 Ω. At resonance, the values of the circuit's inductive reactance and capacitive reactance are each 100 Ω. If the supply voltage is 200 V, calculate (a) the current, (b) the voltage drops across the circuit's resistive component, and (c) across its inductive and capacitive components.

Solution

At resonance, the *only* opposition to current is the resistance of the circuit. So, the resulting current must be:

$$\overline{I} = \frac{\overline{E}}{R} = \frac{200}{10} = 20 \text{ A (Answer a.)}$$

At resonance, the voltage drop across the resistive component will be equal to the supply voltage, i.e. 200 V (Answer b.)

At resonance, the voltage drop across the inductive component is as follows:

$$\overline{U}_L = \overline{I}\, X_L = 20 \times 100 = 2000 \text{ V (Answer c.)}$$

At resonance, the voltage drop across the capacitive component is as follows:

$$\overline{U}_C = \overline{I}\, X_C = 20 \times 100 = 2000 \text{ V (Answer c.)}$$

Yes! The above answers *are* indeed correct! Even though the supply voltage to the circuit is just 200 V, the voltage drop across each of the reactive components is indeed **2000 V**!

If the circuit's resistance was just 1 Ω, instead of 10 Ω, then the current would rise to 200 A, and the voltage drops across the reactive components would become **20 000V**!

Don't forget, these very large voltages are what will appear across the inductive component and the capacitive component, if they are measured *separately*. If we measure the voltage drop across *both* components, then the result would be zero, because the two voltage drops are in antiphase with each other!

So the final thing we can say about resonance is that, *if the resistance of the circuit is low compared with the inductive and capacitive reactances*, then very large voltages can appear across those components —*voltages that are many times the value of the supply voltage!*

> At resonance, if the resistance of the circuit is low compared to the inductive and capacitive reactances, then the individual values of voltage drops, $\overline{U}_L$ and $\overline{U}_C$ can be *many* times larger than the supply voltage.

There are *two* ways of looking at this strange phenomenon. The first is from the **electronics engineer's** point of view; the second is from the **transmission/distribution engineer's** point of view.

Electronics engineers have to deal with signals that are frequently in the microvolt range. Resonance can boost these tiny voltages hundreds of times, providing what is, essentially, 'free amplification' of those signals.

Electricity transmission/distribution engineers, on the other hand, are already dealing with *very* high, and *very* dangerous, voltages. Unintentional resonance could boost these voltages far higher, with disasterous results to system insulation. For example, resonance could occur when a highly-capacitive underground cable feeds a highly-inductive transformer. If the resulting values of capacitive reactance and inductive reactance happen to be such that resonance, or even near-resonance, conditions result, then the resulting voltages could well-exceed the dielectric strengths of the cable/transformer winding insulation, leading to catastrophic failure.

Summary

To summarise what we have learnt about **series resonance**, we can say:

- series resonance occurs when: $X_L = X_C$.
- a series circuit will resonate at a frequency, called its 'resonant frequency' (f_o), which is determined from:

$$f_o = \frac{1}{2\pi\sqrt{LC}}$$

- at resonance, a circuit's impedance will equal its resistance: $Z = R$.
- a circuit's current will reach its maximum value at resonance, and it will be in phase with the supply voltage.
- at resonance, the voltage drop across a circuit's resistive component will equal its supply voltage.
- at resonance, the sum of the voltage drops across a circuit's inductive and capacitive components will be zero.
- at resonance, the individual voltage drops appearing across the inductive or capacitive comonents can be very much higher than the supply voltage.

Exercises

Not all questions are as straightforward as those examples we have seen, so far, and we are often required to think a little harder. For example, let's look at the following example.

1. A coil draws a current of 2 A from a 240-V, 50-Hz, supply, and 4 A from a 100-V d.c. supply. Calculate the coil's resistance and inductance. Hint: remember a 'coil' has both inductance *and* resistance which we always 'separate apart' when drawing its schematic diagram.

   Solution

   This question involves a *coil*. We should realize that a coil has both inductance *and* resistance, so what we have here is a series *R-L* circuit. When supplied from the 100-V d.c. source, the only opposition to current is, of course, the *resistance* of the coil (because *inductive reactance* only occurs in a.c. circuits), so we can easily find that resistance as follows:

   $$R = \frac{E_{dc}}{I_{dc}} = \frac{100}{4} = 25\ \Omega \quad \text{(Answer)}$$

   We can now construct a voltage phasor diagram, from which we can then construct an **impedance diagram** (figure 3.62) for the circuit when it's reconnected to the a.c. supply, as follows…

   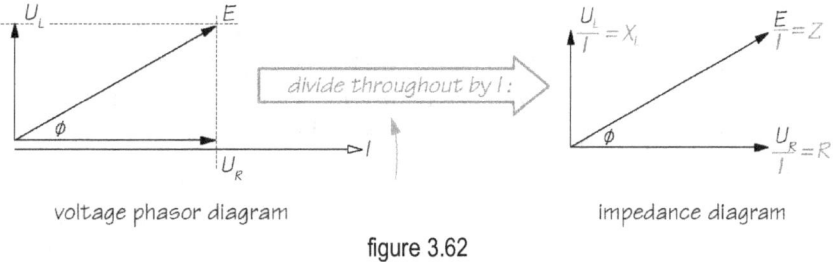

   voltage phasor diagram   impedance diagram

   figure 3.62

   … from which, we can see that we are able to determine the coil's impedance, using the equation:

   $$Z = \frac{E_{ac}}{I_{ac}} = \frac{240}{2} = 120\ \Omega$$

   We can now use Pythagoras's Theorem to determine the inductive reactance…

   $$X_L = \sqrt{Z^2 - R^2} = \sqrt{120^2 - 25^2} = \sqrt{13775} = 117.4\ \Omega$$

Finally, we need to use our 'memorized' equation for inductive reactance, to find the coil's *inductance*:

$$X_L = 2\pi f L$$

$$L = \frac{X_L}{2\pi f} = \frac{117.4}{2\pi \times 50} = 0.374 \text{ H (Answer)}$$

Now, try solving the examples shown below.

2. When a 30-Ω resistor and a capacitor of unknown capacitance are connected in series across a 240-V, 50-Hz, supply, they draw a current of 4.8 A. Calculate the impedance of the circuit, and the capacitance of the capacitor. (Answers: 50 Ω and 79.6 µF)

3. A resistor and an inductor are connected in series across a 250-V, 50-Hz, supply, resulting in a current of 5 A, lagging the supply voltage by an angle of 35°. If the voltage drop measured across the inductor 150 V, calculate the resistance of the resistor, and the resistance and inductance of the inductor. (Answers: 31.8 Ω, 9 Ω, and 0.087 H)

4. A coil, of inductance 5 H and resistance 2 Ω, is connected in series with a variable capacitor, across a 250-V, 50-Hz, supply. What value of capacitance will result in a voltage drop of 270 V appearing across the coil? (Answer: 26.5 µF)

Review what you have learnt

Now that you have completed this chapter, go back and look at the **objectives** listed at the beginning. If you place a question mark at the end of each objective, and ask yourself, *'Can I...'*, then those objectives become **test items**. If you can answer each test item correctly, then you can move on to the next chapter.

Chapter 4

Energy and Power in Alternating Current Circuits

On completion of this chapter, you must be able to

1. define the terms **energy**, **work**, **heat**, and **power**, and specify their SI units of measurement.

2. explain the behaviour of power in
   a. a purely-resistive a.c. circuit.
   b. a purely-inductive a.c. circuit.
   c. a purely-capacitive a.c. circuit.
   d. an *R-L* a.c. circuit.
   e. an *R-C* a.c. circuit.

3. state the relationship between true (or 'active') power, reactive power, and apparent power.

4. state the units of measurement for
   a. true (or 'active') power.
   b. reactive power.
   c. apparent power.

5. change the voltage phasor diagram into a power diagram, and derive equations for true, reactive, and apparent power, for *R-L*, *R-C*, and *R-L-C* circuits.

6. define the term, 'power factor'.

7. solve problems on power in a.c. circuits.

Introduction

In this chapter, we are going to examine the behaviour of **energy** and **power** in alternating-current circuits.

Before we do so, though, we need to remind ourselves of how we define **'energy'**, and how it can be manipulated.

> **Energy** is defined as *'the ability to do work'*.

Energy can be 'manipulated' in either of *two* ways: it can either be ***changed** from one form into another*, or it can be ***transferred** from one body to another*.

When energy is *changed from one form into another*, we say that '**work**' is being done. For example, an electric motor is doing **work** whenever it changes, or converts, electrical energy into kinetic energy.

> **Work** is defined as *'the **conversion** of energy from one form into another'*.

Energy is *transferred between objects* whenever those objects are at different temperatures. We call this process '**heat**', and it describes *'the transfer of energy from the warmer object to the cooler object'*.

> **Heat** is defined as *'the **transfer** of energy from a body at a higher temperature to one at a lower temperature'*.

We say that '**work**' and '**heat**' describe **energy** *'in transit'*, so they are all measured in **joules**.

Finally, we should remind ourselves that '**power**' is defined as *'the rate of doing work'* or *'the rate of heat transfer'*, expressed in joules per second which, in SI, is given the special name: the **watt**.

> **Power** is defined as *'the rate of doing work'* or *'the rate of heat transfer'*.

Behaviour of energy in a.c. circuits

Behaviour of energy in a purely-resistive circuit

Whenever an electric current overcomes the resistance of a metal conductor, it does **work** on that conductor $(W = I^2Rt)$, causing the internal energy (the vibration of its atoms) of the metal to increase. An increase in internal energy is *always* accompanied by an increase in temperature.

If the temperature of the conductor exceeds that of its surroundings, then energy will be transferred *away* from the conductor into its surroundings through **heat transfer**.

This loss of energy through heat transfer *away* from a conductor is *completely irreversible*. That is, we *cannot* transfer that lost energy back into the conductor so that it becomes electrical energy again, and send it back to the supply. Once that energy has gone, it's gone for good, and it ain't coming back!

The *rate* of heat transfer away from a conductor into its surroundings is termed '**power**', and is expressed in watts. For reasons that will shortly become clear, in a purely resistive circuit, it is traditional to refer to this as either '**true power**', '**real power**', or '**active power**' (symbol: P). Throughout this rest of this text, we will stick with the term 'true power'.

Behaviour of energy in a purely-inductive circuit

You will recall that a **purely-inductive circuit** is an 'ideal' circuit, in which there is no resistance.

If a circuit has no resistance, then *no heating can take place.* It might be worth remembering that heat is the consequence of resistance: in other words if there is no resistance, then there can be no heating! So there can be no expenditure or loss of energy, through heat transfer, *away* from a purely-inductive circuit.

However, due to the circuit's **inductance**, the presence of a current creates a *magnetic field*. Because the a.c. current is continuously changing in both magnitude and direction, the resulting magnetic field also varies in magnitude and direction and induces a voltage into the circuit which always acts to *oppose the change in current.*

During the first quarter-cycle, then, the *increasing* current is *opposed* by this induced voltage. Energy is required to overcome this opposition, and this energy is drawn from the supply and stored within the magnetic field. During the second quarter-cycle, as the current *falls*, the direction of the induced voltage reverses, and acts to oppose the reduction of the current —that is, it tries to *sustain* the current. In other words, the energy that was previously *stored* within the magnetic field, is now being *returned* back to the supply.

This process repeats itself during subsequent quarter-cycles, with energy being alternately *stored* in the magnetic field and, then, *released* back to the supply. So, although there *is* energy *conversion* taking place *within* the circuit, there is no net loss of energy *away from* the circuit.

You might like to imagine this process as energy continually *'sloshing back and forth between the supply and the inductive load'*, which is a pretty good description of what is taking place within the circuit!

As always, the *rate* at which this energy 'sloshes' about, is power. But to distinguish it from 'true power', we call it **'reactive power'** (symbol: *Q*). It's also traditional to measure reactive power in **reactive volt amperes** (symbol: **var**), rather than in watts.

> Some textbooks refer to reactive power as *'wattless power'* or *'imaginary power'*. The term, *'imaginary power'*, though, *doesn't* mean it exists only in the mind! In this sense, the word 'imaginary' is used by mathematicians to mean 'quadrature' (i.e. 'at right angles'), in other words it is created by a load current that lags (or, as we shall see, leads) the supply voltage by 90°.

Behaviour of energy in a purely-capacitive circuit

A purely-capacitive circuit is an ideal circuit with no resistance so, again, no heating effect can take place to cause any transfer of energy away from the circuit.

The behaviour of energy in a purely-capacitive circuit is almost identical to its behaviour within a purely-inductive circuit, except that energy from the circuit is being alternately stored in, and returned from, an *electric* field rather than a magnetic field. This electric field, of course, is located within a dielectric. Again, you can imagine this energy as *'sloshing back and forth between the supply and the capacitive load'*.

Once again, although there *is* energy conversion taking place within the circuit, there is no net loss of energy away *from* the circuit.

As with a purely-inductive circuit, the *rate* at which this energy 'sloshes' about, is also called '**reactive power**', which is also measured in **reactive volt amperes** (symbol: **var**) to distinguish it from true power.

Behaviour of energy in *R-L*, *R-C*, and *R-L-C* circuits

In *R-L*, *R-C*, and *R-L-C* circuits, the behaviour of energy must, therefore, be a *combination* of the behaviours of energy we described as happening in purely-resistive and purely-reactive circuits. That is, *some* energy is being irreversably lost from the circuit through *heat transfer* (or, as we shall see, by the *work* done by the load), while *some* energy is being alternately stored, and returned to the supply from, the magnetic or electric fields, in the case of *R-L* and *R-C* circuits, or *both* magnetic *and* electric fields, in the case of an *R-L-C* circuit… with no net loss.

So resistive-reactive circuits have *both* 'true power' *and* 'reactive power' occuring at the same time. Although we *cannot* simply add these two quantities together, the *combination* of the two we call '**apparent power**' (symbol: *S*) and, to distinguish apparent power from true power and reactive power, it is traditional to measure it in **volt amperes***(symbol: **V·A**).

---

*It's worth pointing out that the units, **reactive volt amperes** (**var**) and **volt amperes** (**V·A**), are *traditional*, in order to easily distinguish reactive power, apparent power, and true power from each other. SI doesn't recognise either, and uses the watt to measure all 'forms' of power.

A more-detailed explanation of true power

In the preceding paragraphs we learnt that, for a *purely resistive* circuit, **true power** is *the rate at which energy is irreversibly lost* from the circuit through *heat transfer* to its surroundings. Despite being correct, this explanation is a little simplistic. So let's discuss this in a little more detail, and see if we can come up with a better explanation.

In fact, it's not just the resistive components of a.c. circuits that account for the 'real power' supplied to those circuits. Motors (which are, essentially, *R-L* circuits), for example, do *work* when they convert electrical energy into kinetic energy, in order to drive their mechanical loads.

Remember, '**heat**' and '**work**' are simply two manifestations of *exactly* the same thing: *the expenditure of energy*, the *rate* of which is the '**true power**' of any circuit!

So, in addition to the *rate* at which energy is expended overcoming the electrical *resistance* of a motor circuit, the *rate* at which the motor does **work** when driving its mechanical load, must also represent the '**true power**' of that motor. How do we account for this?

Well, because of its windings, a motor is nothing more than an example of an *R-L* circuit and, as we have learnt ('CI**VIL**'), the motor's supply current will lag the supply voltage by some angle, ϕ.

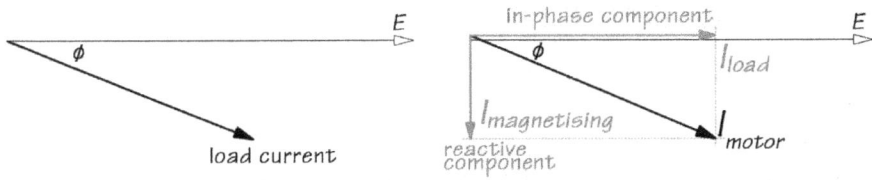

figure 4.1

As can be seen above, in figure 4.1, the motor's (lagging) current (I_{motor}) can actually be **resolved** (broken down) into *two* 'components', called an *'in-phase'* (or *'active'*) component, and a *'reactive'* (meaning 'at 90°') component.

It's the *in-phase component* of the motor's current which is responsible for the **work done** by the motor in driving its mechanical load, and the rate at which this work is done is the 'true power' of the motor load. It's also this in-phase component which accounts for any energy expended by the **resistance** of the motor circuit.

So, it would be more accurate to say that, in an *R-L*, *R-C*, or *R-L-C* circuit, **true power** is the result of the '**in-phase component**' of the load current, rather than simply the *resistance* of those loads.

The **in-phase component** of a load current is responsible for the '**real power**' expended by an *R-L*, *R-C*, or *R-L-C* a.c. circuit.

Summary of true, reactive, and apparent powers

To summarise our explanations for **true power**, **reactive power**, and **apparent power**, we can say that

- the *'in-phase'*, *'real'*, or *'active'* (different names for *exactly* the same thing!) component of a lagging or leading load current is responsible for **true power**, which represents the rate at which energy is expended by the supply —i.e. the rate at which energy is either permanently *lost* through *heat transfer*, or the rate at which *work* is done by a motor, driving its mechanical load, or a combination of the two.

- the *'reactive'*, *'imaginary'* or *'quadrature'* (different names for *exactly* the same thing!) component of the lagging or leading load current is responsible for **reactive power**, which is the rate at which energy is alternately *stored in*, and *returned from*, a magnetic (inductive circuits) or electric field (capacitive circuits), with no net loss from the circuit.

- the load current (i.e. the *vectorial* sum of the true and reactive powers) is responsible for **apparent power**, which is the combination (but *not* the sum) of true power and reactive power.

A.C. power waveforms

Another approach to understanding the behaviour of power in a.c. circuits is to examine their **power waveforms**, which are derived from the corresponding voltage and current waveforms.

Throughout the following explanation, it is conventional to describe the rate at which energy is delivered *from the supply to the load* as being **'positive' power**, while the rate at which energy is *returned from the load to the supply* as being **'negative' power.** So, in this context, the terms 'positive' and 'negative' describe the *direction of flow* of the energy, and have nothing to do with 'electric polarities'.

Purely Resistive Circuits

For a **purely-resistive circuit**, the load current and supply voltage are *in phase* with each other. In figure 4.4, below, the power waveform is constructed by multiplying those values of instantaneous voltage and current, that occur at the same instant in time, over a complete cycle. That is: $p = e\,i$.

To clarify this process, we'll consider just two instants, occurring at points: **A** and **B**:

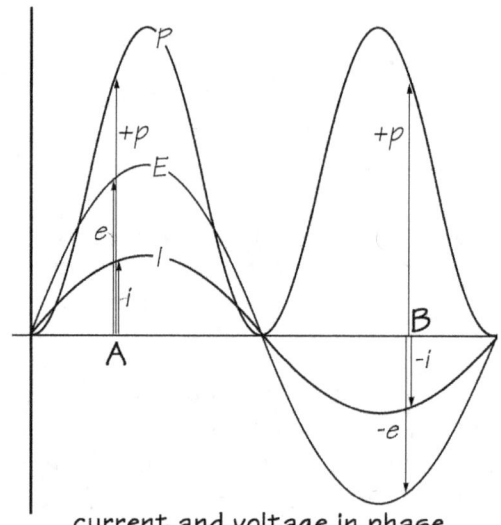

current and voltage in phase

figure 4.4

At point **A**, then, the resulting point on the power waveform is the product of the instantaneous voltage, *e*, and the corresponding instantaneous current, *i*. Both of these are positive, so the corresponding point on the power curve is, therefore, also positive. The power waveform is thus plotted by repeating this process for numerous instantaneous values of current and voltage throughout the first half cycle of the voltage/current waveforms.

At point **B**, the point on the power waveform is, again, the product of the instantaneous voltage and corresponding current at that particular point along the axis. However, this time, both the instantaneous voltage *and* the instantaneous current are *negative*. The resulting point on the power curve is, therefore, *positive*, because the product of two negatives is a positive —that is:

$$-e\,(-i) = +p$$

So, throughout the second half-cycle, we will continue to multiply *negative* instantaneous voltages by *negative* instantaneous currents, which result in *positive* instantaneous power.

The resulting power waveform is illustrated in figure 4.5.

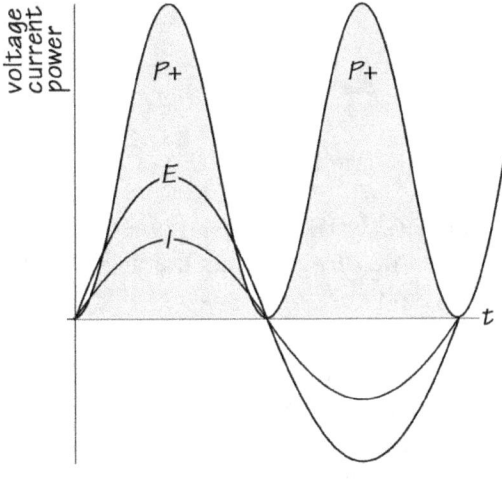

figure 4.5

You will notice that the power waveform is entirely positive. It is called a '**sine-squared' (sine^2) waveform**, meaning that it is a sine wave that varies entirely *above* the horizontal axis (it is also *double* the frequency of the voltage frequency), so that it is entirely *positive* —indicating, by convention, that the energy flow is permanently *from* the supply *towards* the load. The *amount* of power is represented by the grey areas enclosed between the power waveform and the horizontal axis.

> The power output of a quality hi-fi amplifier manufactured by a reputable company is always rated as '**r.m.s. power**', expressed in '**watts (r.m.s.)**'. This is somewhat misleading, because the power output is *not* really being expressed as an r.m.s. value. What it is describing is a power rating that is the product of an *r.m.s. voltage* and an *r.m.s. current*. Less-reputable companies rate their amplifiers as '**total power**', which is the product of a *peak voltage* and a *peak current*, which results in a far-bigger, and more-impressive, figure which is somewhat misleading!

So, for a purely-resistive load, energy is being *expended* by the supply and is being entirely *consumed* by the load or *lost* the the surroundings, through heat transfer, and the rate at which this is happening is termed the **true power** of the circuit, expressed in **watts**.

Some students are confused as to why, when a.c. current reverses direction every half-cyle, the 'direction' of power doesn't *also* reverse every half-cycle too. Remember, when we talk about the *'direction of power'*, what we *really* mean is the *'rate at which energy is flowing in a particular direction'* —the latter being a 'bit of a mouthful', so to speak! In other words, it's just easier and quicker to say *'the direction of power'*! Well, let's consider figure 4.6.

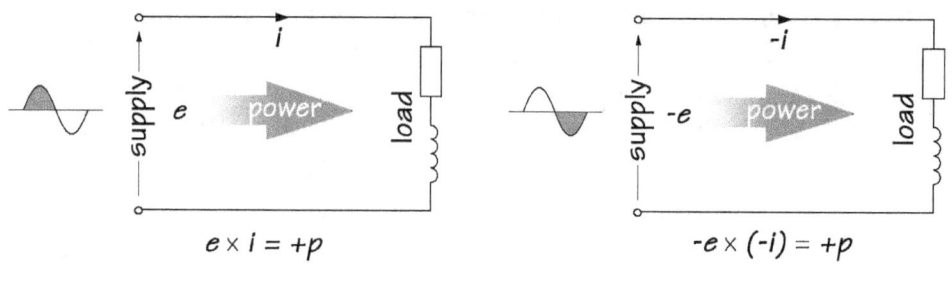

figure 4.6

During the *first* half-cycle (figure 4.6, left), the instantaneous potential differences and currents are both acting in the same *directions* as the assumed sense arrows and, so, their product represents 'positive power': $e \times i = +p$.

During the *second* half-cycle (figure 4.6, right), the instantaneous potential differences and currents are both acting in the *opposite* directions to the assumed sense arrows and, so, are each considered to be negative in direction relative to the first half-cycle. The product of two negatives, of course, is a positive —so, once again, we have 'positive power': $-e \times (-i) = +p$.

Purely-Inductive Circuits

For a purely-inductive circuit, the load current *lags* the supply voltage by 90°. In figure 4.7, the power waveform is, again, constructed by multiplying those values of instantaneous voltage and current, that occur at the same instant of time, throughout a complete cycle. Again, to clarify this process, we'll consider just two instants, at points: **A** and **B**:

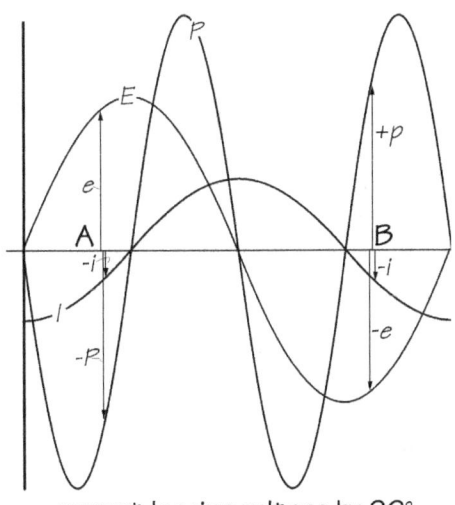

current lagging voltage by 90°

figure 4.7

At point **A**, the point on the power waveform is the product of the instantaneous voltage, e, and the corresponding instantaneous current, $-i$, because at this point along the axis, the instantaneous voltage is positive, whereas the corresponding instantaneous current is negative. The corresponding point on the power curve, therefore, is *negative*:

$$e\,(-i) = -p$$

At point **B**, the point on the power waveform is, again, the product of the instantaneous voltage and current at that particular point along the axis. However, this time, both the instantaneous voltage *and* the instantaneous current are *positive*. The resulting point on the power curve is, therefore, positive, because the product of two negatives is a positive —that is:

$$+e\,(+i) = +p$$

This process is repeated for numerous values of instantaneous voltages and currents throughout a complete cycle of voltage/current, and the resulting power waveform, illustrated in figure 4.8, which is sinusoidal with twice the frequency of the voltage.

You will notice, from the areas enclosed by the power waveform in figure 4.8 above the horizontal axis ('positive' power), are exactly equal to the areas enclosed by the power waveform *below* that axis ('negative' power). So, the rate at which energy is *delivered to* the circuit (the 'positive' power) during each quarter-cycle, and stored in a magnetic field, is exactly *balanced exactly* by the rate at which energy is *returned to* the supply from the collapsing magnetic field (the 'negative' power) during the following quarter-cycle.

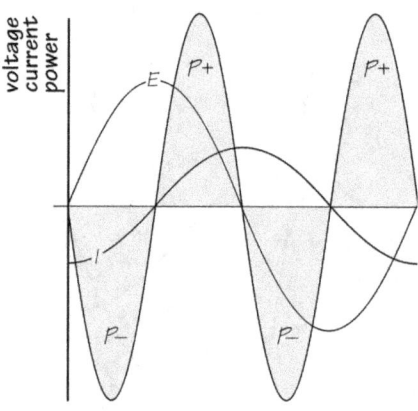

figure 4.8

So, as already explained, the energy in this circuit is being alternately *stored in*, and *returned from*, a magnetic field. No energy is actually being expended by the load or lost to the surroundings through heat transfer, so the 'true power' is zero. On the other hand, the rate at which this energy is being 'sloshed' back and forth represents the **'reactive power'** of the circuit, expressed in **reactive volt amperes**.

Purely-Capacitive Circuits

If we were to draw the power waveform for a purely-capacitive circuit, then we would end up with a similar situation to that of a purely-inductive circuit. That is, the amount of 'positive' power would be exactly balanced by the amount of 'negative' power. So, once again, the rate at which energy is being *delivered* every quarter-cycle to the load and stored in an electric field is balanced exactly by the rate at which energy is being *returned* to the supply when that field collapses during the following quarter-cycle.

So, as for a purely-inductive circuit, with no energy is being consumed by the load or lost through heat transfer, the 'true power' is zero. But the rate at which energy is being 'sloshed' back and forth represents the **'reactive power'** of the circuit.

Power in resistive-reactive circuits

In the chapter on *A.C. Series Circuits*, we learnt that 'real' circuits are a *combination* of resistance, inductance, and/or capacitance. So, in a 'real circuit' —i.e. a resistive-reactive circuit— then

- some energy is being *permanently lost* due to the resistive component of the circuit, through heat transfer and/or the work done by loads such as motors while, at the same time…

- some energy is being alternately stored, and returned ('sloshing between') to the load from the magnetic and/or electric fields associated with the circuit's *reactive (inductive and/or capacitive) components*.

In 'real' circuits, therefore, there exists both **true-power** (the rate at which energy is *permanently* lost) *and* **reactive-power** (the rate at which energy is continually stored in, and returned from, magnetic or electric fields.

In the power waveform diagram for an *R-L* circuit (e.g. an electric motor), figure 4.9, the current lags the supply voltage by some phase-angle, ϕ. You will notice that the amount of 'positive' power *above* the horizontal axis, is *larger* than the amount of 'negative' power, *below* that axis. This means that the rate of energy transfer *to* the load is *greater* than the rate of energy transfer (temporarily stored in the magnetic field) from the load back to the supply.

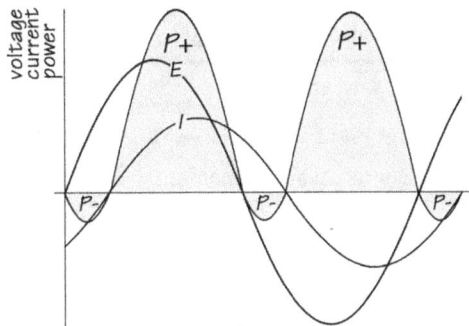

figure 4.9

To summarise, we can say that **true power** is the rate at which energy is *lost* through heat transfer or by the work done by a load such as a motor, whereas **reactive power** is the rate at which energy must be alternately supplied/returned in order to *sustain the magnetic and/or electric fields*.

> We should *not* assume that reactive power is unimportant; in fact, reactive power is *essential* to the operation of an electrical transmission and distribution system, in order to maintain magnetic/electric fields, and to maintain the system voltages required to 'push' the energy demanded by loads along the transmission lines.

Apparent-power, true-power, and reactive-power

Now let's move on to examine the mathematical relationship between **apparent power**, **true power**, and **reactive power**.

As we have already learnt, **apparent power** is the *combination* of a circuit's true power and reactive power. However, the relationship between apparent-power, true-power, and reactive-power is *not* a simple algebraic relationship but, rather, a *vectorial* relationship, as we will see shortly.

Units of Measurement

Whether we are discussing **apparent-power, true-power,** *or* **reactive-power**, we should bear in mind that 'power' is *always* the *rate* at which energy is being moved around —*regardless* of whether that work is reversible or irreversible, useful or useless! So, there is absolutely no technical reason, therefore, why each of these quantities shouldn't be measured using the *same* unit of measurement —i.e. the **watt** (symbol: **W**). In fact, that is precisely what SI does! SI doesn't acknowledge the following units.

Traditionally, however, *in order to clearly distinguish between each of these quantities*, different units of measurement have been allocated to them, and it is unlikely we will ever see them being replaced by the watt! In fact, using these distinctive units of measurement is very useful, as there can then be absolutely no doubt as to which 'form' of power is then being referred to. This is summarized in table 4.1:

quantity:	symbol:	unit of measurement:	symbol:
apparent-power	S	volt ampere	V·A
true-power	P	watt	W
reactive-power	Q	reactive volt ampere	var

table 4.1

Power in a series R-L circuit

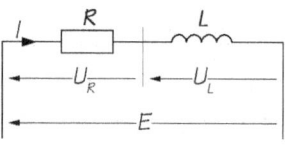

figure 4.10

You will recall that to convert a phasor diagram into an *impedance diagram*, we divided throughout by the reference phasor, $\overline{I}$. Doing this generated various useful equations for *resistance, inductive reactance,* and *impedance,* as well as for the *phase-angle*.

Well, by *multiplying* the phasor diagram by the reference phasor, $\overline{I}$, we can produce a **power diagram** and, at the same time, generate various equations that will allow us to determine the circuit's **true power, reactive power, apparent power,** and **power factor** (more on this, later).

Step 1. We start by drawing the voltage phasor diagram for the circuit, just as we did in the chapter on *A.C. Series Circuits* (figure 4.11).

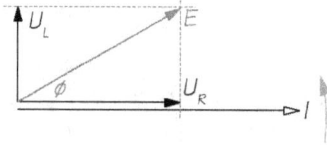

figure 4.11

Step 2: *Multiply* throughout by the reference phasor, $\overline{I}$ (figure 4.12).

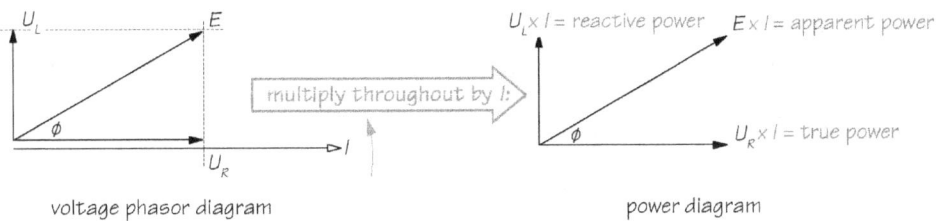

figure 4.12

So, by simply *multiplying the voltage phasor diagram by $\overline{I}$*, we have created a **power diagram**, from which the following equations have been derived:

$$\boxed{\text{Apparent Power} = \overline{I}\,\overline{E}} \qquad \boxed{\text{True Power} = \overline{I}\,\overline{U}_R} \qquad \boxed{\text{Reactive Power} = \overline{I}\,\overline{U}_L}$$

If we know *any two* out of these three, then *we can find the third*, by applying *Pythagoras' Theorem*. For example, to find apparent power in terms of true and reactive power (figure 4.13).

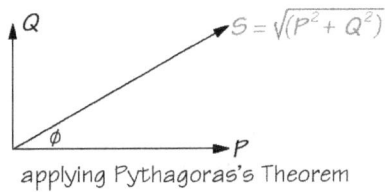

applying Pythagoras's Theorem

figure 4.13

$$(\text{Apparent Power})^2 = (\text{True Power})^2 + (\text{Reactive Power})^2$$

$$S = \sqrt{P^2 + Q^2}$$

To demonstrate further how useful phasor diagrams are in generating equations for power, let's look at another example of using this technique.

We know that power in a resistive circuit can be obtained from $P = \overline{I}^2 R$, so if we redraw the impedance diagram, and multiply throughout by $\overline{I}^2$, we'll end up with another version of the power diagram, as shown in figure 4.14.

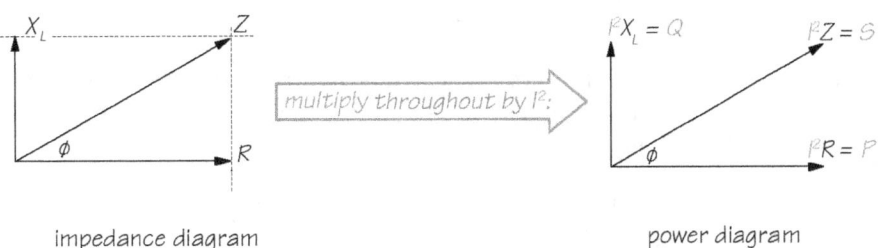

figure 4.14

Thus generating the following additional equations for power!

$$\text{Apparent Power} = \overline{I}^2 Z \qquad TruePower = \overline{I}^2 R \qquad \text{Reactive Power} = \overline{I}^2 X_L$$

> **Note!**
> Hopefully, you are now beginning to realise how important it is to be able to construct a voltage phasor diagram and to be able to change this into an impedance diagram and into a power diagram. These diagrams will generate *all* the equations that you will ever need to solve a.c. problems, *without the need to commit any of these equations to memory!*

Power factor

In a d.c. circuit, *regardless of the type of load*, we can determine the power of that load simply by multiplying together the readings of a voltmeter and an ammeter.

In a resisitive-reactive *a.c.* circuit, however, the product of the supply voltage and the load current, as we have learnt, gives us the **apparent power** of the load, *not* its true power.

To determine its **true power** we must use a *wattmeter*, which is designed specifically to measure true power by monitoring the supply voltage together with the *in-phase (resistive) component* of the load current (figure 4.15).

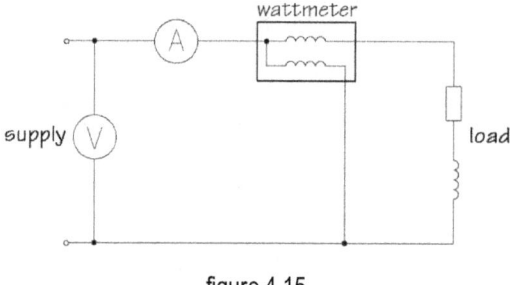

figure 4.15

The *ratio of a load's true power to its apparent power* is *very* important, and is called the '**power factor**' of the load:

$$\text{power factor} = \frac{\text{true power}}{\text{apparent power}}$$

If we re-examine the power diagram (figure 4.13), derived from the voltage phasor diagram, it should be obvious that the ratio of true power to apparent power (adjacent over hypotenuse) is the **cosine** of the phase angle:

$$\text{power factor} = \frac{\text{true power}}{\text{apparent power}} = \frac{\text{adjacent}}{\text{hypotenuse}} = \cos\phi$$

So an *alternative* definition for power factor is that it is '*the cosine of the load's phase angle*'.

Power factor is usually expressed as a *per-unit* value (e.g. 0.85), although it may still occasionally be seen expressed as a *percentage* value (e.g. 85%).

A circuit's power-factor *must* also be specified as either '*leading*' or '*lagging*' (e.g '0.85 lagging'). These terms always refer to *where the **load current** phasor lies in relation to the supply-voltage phasor* (never the other way around!). **Resistive-inductive** circuits, therefore, *always* have '**lagging**' **power-factors**, while **resistive-capacitive** circuits *always* have '**leading**' **power-factors**.

It might prove helpful to think of power factor in terms of the following mechanical analogy.

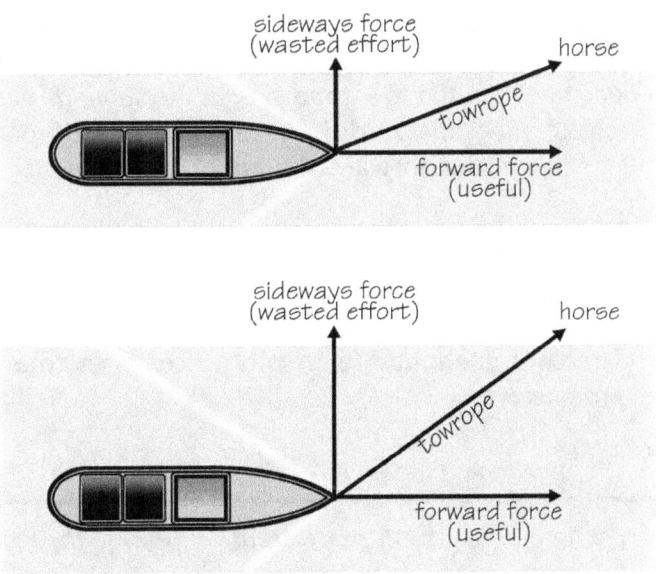

figure 4.16

The two diagrams in figure 4.16, represent a plan view of a barge which is being towed along a canal by a horse, walking along the canal's towpath.

The force in the tow-rope can be resolved into two forces: one *in the forward direction*, and the other *towards the canal's bank*.

Of these two resolved forces, it is the one acting in the forward direction of the barge's motion that contributes the *useful* force, as it's acting to pull the barge in the desired direction. On the other hand, the resolved force acting towards the bank (quadrature force) contributes nothing to the forward motion to the barge. Yet neither force can exist without the other!

In the lower illustration, the canal is wider, and the barge is further away from the bank, so the tow-rope assumes a somewhat greater angle than before.

If we, again, resolve the force in the tow-rope (assuming that the same forward resolved force is required to keep the barge moving at the same velocity), we notice that the resolved force towards the canal bank is *greater* than it was before.

In this new situation, *in order to provide the same amount of forward force*, a greater force *has to be made available in the direction of the bank.*

This situation, may be likened to the power supplied to a resistive-reactive circuit. The **true power** is equivalent to *the forward force acting on the barge*; the **reactive power** is equivalent to the *'wasted' force acting towards the canal bank* (which, despite being *apparently* 'wasted', is still *essential* to the behaviour of the barge). The greater the phase-angle (equivalent to the angle between the forward force on the barge and the force in the tow-rope), the more 'apparently-wasted' (but, none-the-less *essential*!) reactive power that must be provided!

By expressing the angle between the tow-rope and the forward-resolved force in terms of its cosine, the 'efficiency' of the arrangement is indicated. The maximum efficiency (×1 or ×100%) occurs when the tow-rope acts in the *same direction as the necessary forward force* ($\cos 0° = 1$); and zero efficiency (×0 or 0%) occurs when tow-rope is at 90° ($\cos 90° = 0$) to the forward direction of the barge —i.e. no forward motion whatsoever!

While this analogy is useful, for **power factor** 'efficiency' *isn't* really the appropriate term to describe what it represents. This is because we are really comparing *different quantities*: i.e. *apparent* power with *true* power. To truly measure 'efficiency', we would have to compare the *same* quantities (e.g. true power with true power, or apparent power with apparent power).

So, it is more accurate to say that...

> **Power Factor** is *'the percentage of apparent power that represents the rate of doing real work'.*

In other words,

- when the power factor is unity (1), the apparent power is equal to the true power.
- when the power factor is zero (0), the apparent power is equal to the reactive power.

For most practical circuits, the power factor lies somewhere in between these two extremes —usually closer to unity than to zero.

Power in a series R-C circuit

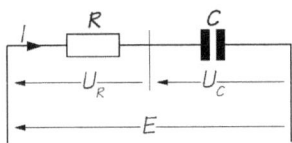

figure 4.17

As with the *R-L* circuit, described earlier, to convert a phasor diagram into a *power diagram*, we simply **multiply the phasor diagram by the reference phasor**, $\overline{I}$, and generate equations that will allow us to determine the circuit's true power, reactive power, apparent power, and power factor.

Step 1. Draw the voltage phasor diagram for the above circuit, just as we did in the chapter on *A.C. Series Circuits* (figure 4.18).

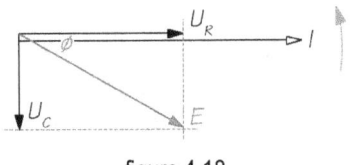

figure 4.18

Step 2: *Multiply* throughout by the reference phasor, $\overline{I}$ (figure 4.19).

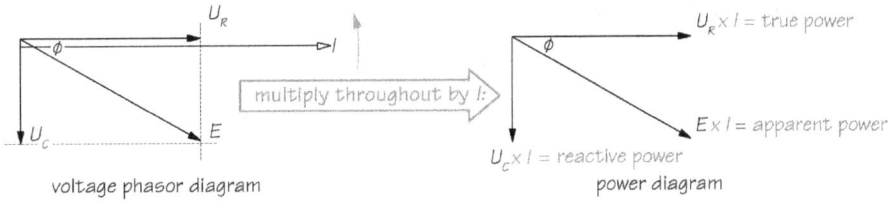

figure 4.19

So, by simply *multiplying the voltage phasor diagram by $\overline{I}$*, we have created a **power diagram**, from which the following equations have been generated:

$$\text{Apparent Power} = \overline{I}\,\overline{E} \qquad \text{TruePower} = \overline{I}\,\overline{U}_R \qquad \text{Reactive Power} = \overline{I}\,\overline{U}_C$$

If we know *any two* out of these three, then *we can find the third*, by applying *Pythagoras' Theorem* (figure 4.20). For example:

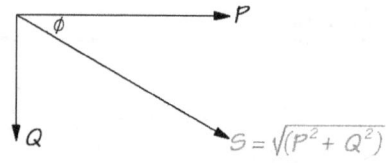
applying Pythagoras's Theorem

figure 4.20

$$(\text{Apparent Power P})^2 = (\text{True Power})^2 + (\text{Reactive Power})^2$$

$$S = \sqrt{P^2 + Q^2}$$

So, once again, the ability to draw a power diagram *saves us from having to remember numerous equations*, as they can *all* be generated from the power diagram, by applying simple geometry (Pythagoras' Theorem) or simple trigonometry (the cosine ratio)! Let's look at what else we can derive from the power diagram:

For example, if we know the *Apparent Power* and the *True Power*, then we could rearrange Pythagoras' Theorem, as follows:

$$\text{Reactive Power} = \sqrt{(\text{Apparent Power})^2 - (\text{True Power})^2}$$

To find the **power factor**, we simply need to find the cosine of the phase angle:

$$\cos\phi = \frac{\text{adjacent}}{\text{hypotenuse}} = \frac{\text{True Power}}{\text{Apparent Power}}$$

This time, we *always* describe that the power factor as **leading**, because, by definition, the *current leads the supply voltage*.

Again, we can manipulate this equation, to find another equation for the True Power of a circuit:

$$\text{since } \cos\phi = \frac{\text{True Power}}{\text{Apparent Power}}$$

then True Power = Apparent Power $\times \cos\phi$

This gives us a very important equation:

$$\boxed{\text{True Power} = (\overline{E}\overline{I})\cos\phi}$$

As was the case for the series *R-L* circuit, we can also convert an *R-C* impedance diagram into a power diagram, by simply multiplying throughout by I^2 (figure 4.21).

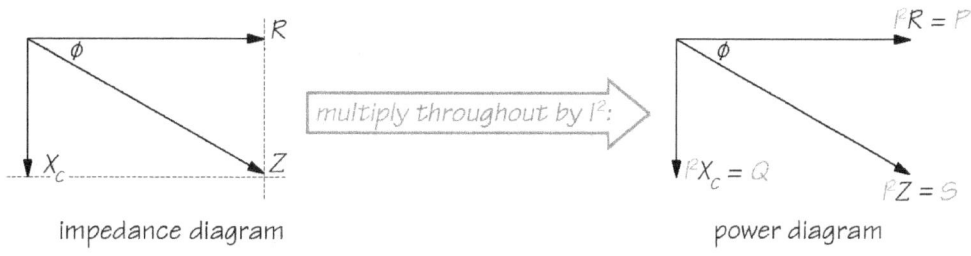

figure 4.21

Thus generating the following additional equations for power:

Apparent Power $= \overline{I}^2 Z$ TruePower $= \overline{I}^2 R$ Reactive Power $= \overline{I}^2 X_C$

Note!

If you learn *nothing else* from this Unit, you *must* learn **the importance of being able to draw a voltage phasor diagram and to be able to convert it into a power diagram**, for this will generate *all* the equations you will *ever* need to know in order to solve a.c. power problems.

This will save you from *ever* needing to memorise these equations!

Worked Example 1

An inductor, of inductance 16 mH and resistance 2 Ω is connected across a 24-V, 50 Hz, supply. Calculate each of the following:

a. Inductive Reactance.
b. Impedance.
c. Current.
d. Apparent Power.
e. True Power.
f. Reactive Power.
g. Power Factor.

Solution

As with any problem, always start by sketching the circuit and inserting all the values given in the problem (figure 4.22). . You notice that this circuit consists of an inductor; you are given its inductance and resistance, so you can assume that it is equivalent to a *series R-L* circuit.

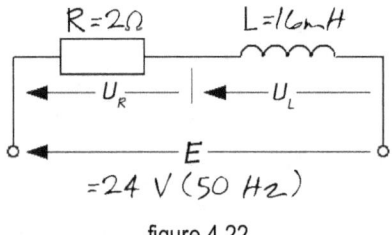

figure 4.22

Next, sketch the corresponding voltage phasor diagram (figure 4.23).

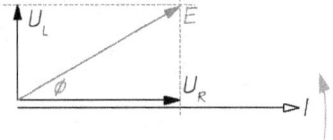

figure 4.23

Next, convert this into an impedance diagram, by dividing throughout by $\overline{I}$ (figure 4.24).

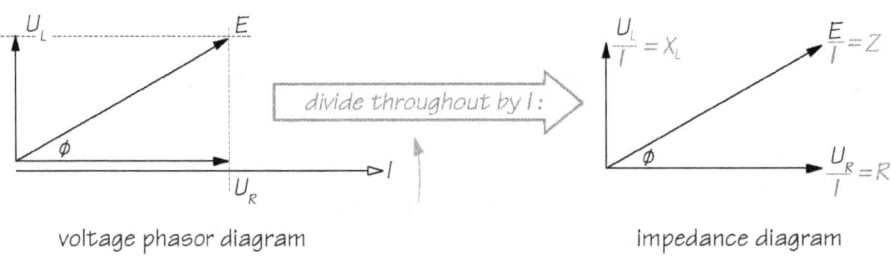

figure 4.24

a. None of the generated equations allow us to determine the inductive reactance, so we must fall back on the basic equation that we have committed to memory:

$$X_L = 2\pi f = 2\pi \times 50 \times (16 \times 10^{-3}) = 5.03\,\Omega \text{ (Answer a.)}$$

b. $Z = \sqrt{R^2 + X_L^2} = \sqrt{2^2 \times 5.03^2} = \sqrt{29.30} = 5.41\,\Omega$ (Answer b.)

c. To find the current, we can use the following equation generated by the impedance diagram:

$$\overline{I} = \frac{E}{Z} = \frac{24}{5.41} = 4.44 \text{ A (Answer c.)}$$

For the rest of this problem, redraw the impedance diagram, and multiply by $\overline{I}^2$ to create a power diagram. Then, we can utilise the equations generated by the power diagram to solve the remaining parts of the problem (figure 4.25).

figure 4.25

d. Apparent Power $= \overline{I}^2 Z$

$= 4.44^2 \times 5.41 = 106.65$ V·A (Answer d.)

e. True Power $= \overline{I}^2 R = 4.44^2 \times 2 = 39.43$ W (Answer e.)

f. Reactive Power $= \overline{I}^2 X_L$

$= 4.44^2 \times 5.02 = 98.96$ var (Answer f.)

g. $\text{PowerFactor} = \dfrac{\text{True Power}}{\text{Apparent Power}}$

$= \dfrac{39.43}{106.65} = 0.3697$ lagging (Answer g.)

*('Lagging', because it is an **inductive** circuit, and current always lags the supply voltage in an inductive circuit)*

Exercises

Once again, not all questions are as straightforward as those examples we have seen, so far, and we are often required to think a little harder. For example, let's look at the following example.

1. When a 30-Ω resistor and a capacitor of unknown capacitance are connected in series across a 240-V, 50-Hz, supply, the resulting current is found to be 4.8 A. Find the true power and the power-factor of the circuit.

    ### Solution

    As we have learnt, true power is *only* developed in the *resistive* component, not in the capacitive component of any circuit. In this example, we know the value of the resistance and the current through it, so it is very easy to find the true power:

    $$P = \overline{I}^2 R = 4.8^2 \times 30 = 691 \text{ W (Answer)}$$

    If we construct a power diagram from a voltage phasor diagram, it's also very easy to find the apparent power of the circuit:

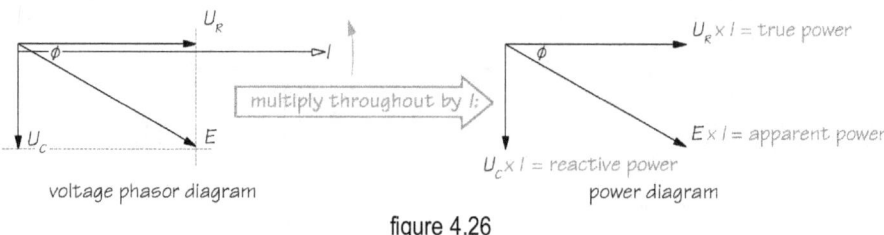

figure 4.26

$$S = \overline{E}\,\overline{I} = 240 \times 4.8 = 1152 \text{ V} \cdot \text{A}$$

And, since *power factor* is the ratio of *true power* to *apparent power*:

$$\text{power factor} = \cos\phi = \frac{P}{S} = \frac{691}{1152} = 0.6 \text{ leading (Answer)}$$

'Leading' because, of course, it's a capacitive circuit.

Now, try solving the examples shown below.

2. A capacitor is found to have a reactive power of 50 var when connected across a 120-V, 60-Hz, supply. Calculate the value of the capacitor. (Answer: 9.2 µF)

3. A 20-Ω resistor is connected in series with a coil across a 100-V, 50-Hz, supply. A voltmeter connected across the resistor reads 46 V and, when connected across the coil, reads 67.5 V. Calculate the true power of the circuit, the reactive power of the coil, and the power factor of the circuit. (Answers: 83.23 W, 131 var, 0.822). Hint: remember that the coil has resistance as well as inductance!

4. A coil was found to have a true power of 0.7 kW and a reactive power of 1.8 kvar when connected across a 240-V, 50-Hz, supply. If that coil is now connected in series with a capacitor of capacitance 80 µF, across the same supply, calculate the resulting current, power factor, and true power. (Answers: 15 A, 0.68 leading, 2440 W)

Review what you have learnt

Now that you have completed this chapter, go back and look at the **objectives** listed at the beginning. If you place a question mark at the end of each objective, and ask yourself, *'Can I...'*, then those objectives become **test items**. If you can answer each test item correctly, then you can move on to the next chapter.

Annexe 1
Summary of Equations for Series A.C. Circuits

In the preceding two chapters, we learnt how to construct phasor diagrams for a **series a.c. circuits**. By constructing a phasor diagram, solving an 'electrical' problem becomes a simple matter of applying either *basic geometry* (Pythagoras's Theorem) or *basic trigonometry* (sine, cosine, or tangent ratios) to simple right-angled triangles.

Understanding this concept is of vital importance in understanding the behaviour of alternating current. So much so that, here, we are *not* going to learn anything new, but we are going to summarise all that we —hopefully— will have learnt from this chapter.

Voltage phasor diagrams

For example, by applying **Pythagoras's Theorem** to each of the following voltage phasor diagrams, which are really just simple **right-angled triangles**, the following relationships are revealed:

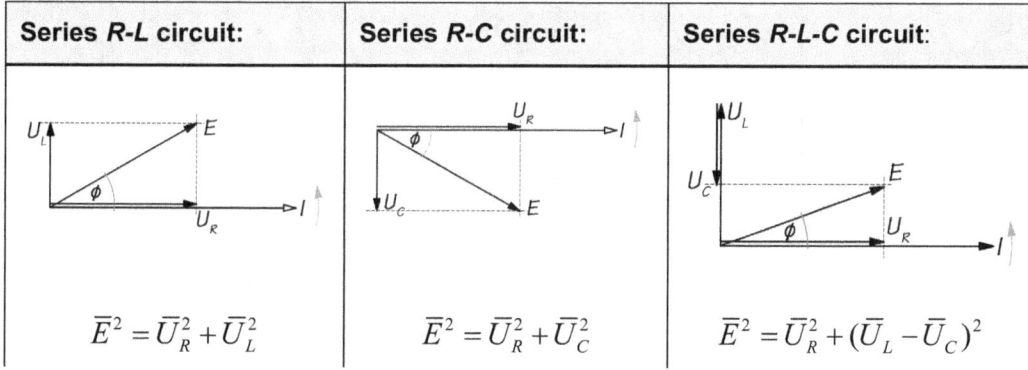

Series *R-L* circuit:	Series *R-C* circuit:	Series *R-L-C* circuit:
$\overline{E}^2 = \overline{U}_R^2 + \overline{U}_L^2$	$\overline{E}^2 = \overline{U}_R^2 + \overline{U}_C^2$	$\overline{E}^2 = \overline{U}_R^2 + (\overline{U}_L - \overline{U}_C)^2$

From which we can derive the following equations:

$\overline{E} = \sqrt{\overline{U}_R^2 + \overline{U}_L^2}$	$\overline{E} = \sqrt{\overline{U}_R^2 + \overline{U}_C^2}$	$\overline{E} = \sqrt{\overline{U}_R^2 + (\overline{U}_L - \overline{U}_C)^2}$
$\overline{U}_R = \sqrt{\overline{E}^2 - \overline{U}_L^2}$	$\overline{U}_R = \sqrt{\overline{E}^2 - \overline{U}_C^2}$	$\overline{U}_R = \sqrt{\overline{E} - (\overline{U}_L - \overline{U}_C)^2}$
$\overline{U}_L = \sqrt{\overline{E}^2 - \overline{U}_R^2}$	$\overline{U}_C = \sqrt{\overline{E}^2 - \overline{U}_R^2}$	$(\overline{U}_L - \overline{U}_C) = \sqrt{\overline{E} - \overline{U}_R^2}$

We can also determine the phase angle for each circuit, as follows:

$\angle \phi = \cos^{-1}\left(\dfrac{\overline{U}_R}{\overline{E}}\right)$	$\angle \phi = \cos^{-1}\left(\dfrac{\overline{U}_R}{\overline{E}}\right)$	$\angle \phi = \cos^{-1}\left(\dfrac{\overline{U}_R}{\overline{E}}\right)$

Impedance Diagrams

We can convert each of the above *voltage* phasor diagrams into a corresponding **impedance diagram**, by simply *dividing throughout by the reference phasor*, $\overline{I}$, as follows:

Series *R-L* circuit:	Series *R-C* circuit:	Series *R-L-C* circuit:

From which we get the following equations:

$Z = \dfrac{\overline{E}}{\overline{I}}$	$Z = \dfrac{\overline{E}}{\overline{I}}$	$Z = \dfrac{\overline{E}}{\overline{I}}$
$R = \dfrac{\overline{U}_R}{\overline{I}}$	$R = \dfrac{\overline{U}_R}{\overline{I}}$	$R = \dfrac{\overline{U}_R}{\overline{I}}$
$X_L = \dfrac{\overline{U}_L}{\overline{I}}$	$X_C = \dfrac{\overline{U}_C}{\overline{I}}$	$X_L = \dfrac{\overline{U}_L}{\overline{I}}$
		$X_C = \dfrac{\overline{U}_C}{\overline{I}}$
		$X = \dfrac{(\overline{U}_L - \overline{U}_C)}{\overline{I}}$

Applying **Pythagoras's Theorem** to each impedance diagram:

$$Z^2 = R^2 + X_L^2 \qquad Z^2 = R^2 + X_C^2 \qquad Z^2 = R^2 + (X_L - X_C)^2$$

From which we can then find the values of Z, X_L, and X_C:

$Z = \sqrt{R^2 + X_L^2}$	$Z = \sqrt{R^2 + X_C^2}$	$Z = \sqrt{R^2 + (X_L - X_C)^2}$
$R = \sqrt{Z^2 - X_L^2}$	$R = \sqrt{Z^2 - X_C^2}$	$R = \sqrt{Z^2 + (X_L - X_C)^2}$
$X_L = \sqrt{Z^2 - R^2}$	$X_C = \sqrt{Z^2 - R^2}$	$(X_L - X_C) = \sqrt{Z^2 - R^2}$

We can also determine the phase angle for each circuit, as follows:

| $\angle\phi = \cos^{-1}\left(\dfrac{R}{Z}\right)$ | $\angle\phi = \cos^{-1}\left(\dfrac{R}{Z}\right)$ | $\angle\phi = \cos^{-1}\left(\dfrac{R}{Z}\right)$ |

Power Diagrams

We can convert any *voltage* phasor diagram into a corresponding **power diagram**, simply by *multiplying throughout by the reference phasor,* $\overline{I}$, as follows.

Remember, '**apparent power**' *(S)* is expressed in volt amperes, '**true power**' *(P)* in watts, and reactive power *(Q)* in reactive volt amperes.

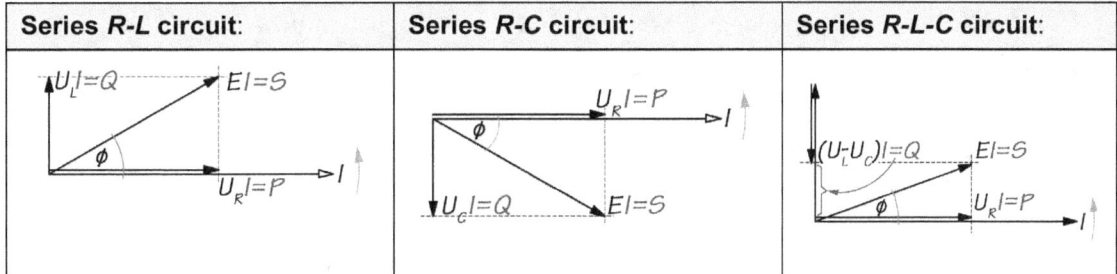

| Series *R-L* circuit: | Series *R-C* circuit: | Series *R-L-C* circuit: |

From which we get the following equations:

apparent power, $S = \overline{E}\overline{I}$	apparent power, $S = \overline{E}\overline{I}$	apparent power, $S = \overline{E}\overline{I}$
true power, $P = \overline{U}_R \overline{I}$	true power, $P = \overline{U}_R \overline{I}$	true power, $P = \overline{U}_R \overline{I}$
reactive power, $Q = \overline{U}_L \overline{I}$	reactive power, $Q = \overline{U}_C \overline{I}$	$Q = (\overline{U}_L - \overline{U}_C)\overline{I}$

Applying **Pythagoras's Theorem** to each power diagram:

| $S^2 = P^2 + Q^2$ | $S^2 = P^2 + Q^2$ | $S^2 = P^2 + Q^2$ |

From which, we can find the values of apparent power *(S)*, true power *(P)*, and reactive power *(Q)*:

$S = \sqrt{P^2 + Q^2}$	$S = \sqrt{P^2 + Q^2}$	$S = \sqrt{P^2 + Q^2}$
$P = \sqrt{S^2 - Q^2}$	$P = \sqrt{S^2 - Q^2}$	$P = \sqrt{S^2 - Q^2}$
$Q = \sqrt{S^2 - P^2}$	$Q = \sqrt{S^2 - P^2}$	$Q = \sqrt{S^2 - P^2}$

We can also convert any *impedance* diagram into a **power diagram**, by *multiplying throughout by the square of the current* $(\overline{I}^2)$, as follows.

Series *R-L* circuit:	Series *R-C* circuit:	Series *R-C-L* circuit:
$X_L I^2 = Q$ $ZI^2 = S$ $RI^2 = P$	$RI^2 = S$ $X_C I^2 = Q$ $ZI^2 = P$	$ZI^2 = S$ $(X_L - X_C)I^2 = Q$ $RI^2 = P$

From which we get the following equations:

apparent power, $S = \overline{I}^2 Z$ true power, $P = \overline{I}^2 R$ reactive power, $Q = \overline{I}^2 X_L$	apparent power, $S = \overline{I}^2 Z$ true power, $P = \overline{I}^2 R$ reactive power, $Q = \overline{I}^2 X_C$	apparent power, $S = \overline{I}^2 Z$ true power, $P = \overline{I}^2 R$ $Q = \overline{I}^2 (X_L - X_C)$

Chapter 5
Parallel Alternating-Current Circuits

On completion of this Chapter, you must be able to

1. draw a current phasor diagram for a parallel:
    a. resistive-inductive *(R-L)* circuit.
    b. resistive-capacitive *(R-C)* circuit.
    c. resistive-inductive-capacitive *(R-L-C)* circuit.

2. derive an admittance diagram, and derive expressions for conductance, susceptance, admittance, and phase-angle for a parallel:
    a. resistive-inductive *(R-L)* circuit.
    b. resistive-capacitive *(R-C)* circuit.
    c. resistive-inductive-capacitive *(R-L-C)* circuit.

3. derive a power diagram, and derive expressions for true power, reactive power, apparent power, and power factor for a parallel:
    a. resistive-inductive *(R-L)* circuit.
    b. resistive-capacitive *(R-C)* circuit.
    c. resistive-inductive-capacitive *(R-L-C)* circuit.

4. solve problems on parallel *R-L*, *R-C*, and *R-L-C* circuits.

Introduction

Just as it was with the previous two chapters, the *key* to understanding and solving **parallel a.c. circuits** is the ability to be able to draw phasor diagrams from which *practically every equation you will ever need can be derived* —*providing you know how to use Pythagoras's Theorem and basic trigonometry.*

So, once again, you are urged *not* to waste your time trying to memorise all the various equations that you are about to meet!

We have learnt how *all* 'real' a.c. circuits exhibit combinations of **resistance** (symbol: *R*), **inductance** (symbol: *L*), and **capacitance** (symbol: *C*).

We also learnt that 'real' circuits are relatively complicated because they contain *combinations of resistance, inductance, and capacitance* and, in order to understand the behaviour of a.c. circuits, it is necessary to start by first considering how 'ideal' circuits would behave. We described these 'ideal' circuits as **purely resistive, purely inductive**, and **purely capacitive**, and we discovered that

a. in a **purely-resistive** circuit, *the current and voltage are in phase* with each other.

b. in a **purely-inductive** circuit, *the current lags the voltage by 90°*.

c. in a **purely-capacitive** circuit, *the current leads the voltage by 90°*.

Again, to help us remember these *very* important relationships, we can use the mnemonic, '**CIVIL**' (figure 5.1), in which '**C**' stands for 'capacitive circuit', and '**L**' stands for '**inductive** circuit':

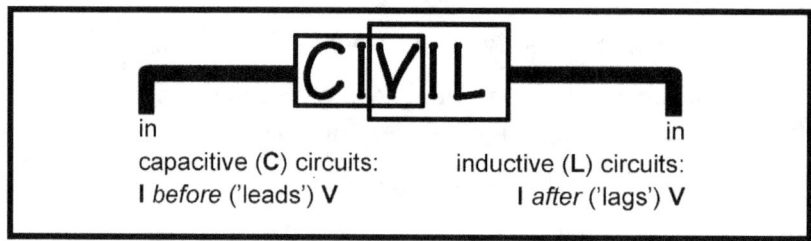

figure 5.1

Or, if we prefer, we can use the mnemonic, **ELI** the **ICE**man.

Finally, we discovered that the opposition to current in a purely-resistive circuit is called **resistance** (R), the opposition to current in a purely-inductive circuit is called **inductive reactance** (X_L), and that the opposition to current in a purely-capacitive circuit is called **capacitive reactance** (X_C), where:

$$X_L = 2\pi f L \qquad X_C = \frac{1}{2\pi f C}$$

where:
X_L = inductive reactance (ohms)
X_C = capacitive reactance (ohms)
f = supply frequency (hertz)
L = inductance (henrys)
C = capacitance (farads)

Unlike *all* the other equations that we will need to use for solving a.c. circuits, these two equations, unfortunately, *have to be committed to memory* because their derivation is beyond the scope of this book.

All other equations can be derived by learning to represent circuits using phasor diagrams and their derivatives (e.g. impedance diagrams and power diagrams). So if we learn how to draw these diagrams, then we will have absolutely no need to remember *any* the other various equations.

In this chapter we will be examining **parallel a.c. circuits**. Specifically, **parallel R-L**, **parallel R-C**, and **parallel R-L-C** circuits.

Again, a reminder that the circuit symbols used throughout this chapter represent the *quan*tities resist*ance*, induct*ance*, and capacit*ance* —***not*** the circuit components: resist*ors*, induct*ors*, and capacit*ors*.

Parallel R-L circuits

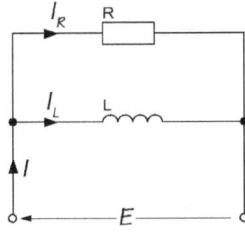

figure 5.2

When a potential difference $(\overline{E})$ is applied across a parallel circuit (figure 5.2), *that potential difference is common to each branch*. For an *R-L* parallel circuit, the resistive branch will then draw a current $\overline{I}_R$, and the inductive branch will draw a current $\overline{I}_L$.

In accordance with Kirchhoff's Current Law, the supply current $(\overline{I})$ will then be the *sum* of the two branch currents. However, as we learnt in the chapter on **A.C. Series Circuits**, because these currents are not in phase with each other, *we must add them vectorially*.

So, let's examine this a little more closely, by drawing the phasor diagram for the circuit.

Drawing the Phasor Diagram
Step 1*:*
In a parallel circuit, the **supply voltage** $(\overline{E})$ is common to each branch and, so, *voltage is always chosen as the **reference phasor***.

> **Important!** In **parallel** circuits, the supply voltage is common to each branch of that circuit, so **voltage** is *always* chosen as the reference phasor.

The reference phasor is *always drawn first, and always along the horizontal positive axis* (i.e. horizontally and to the right), and it's also normally drawn fairly long in order to distinguish it from the others. In figure 5.3, we have also given the reference phasor an outline, rather than a solid, arrow head although this is not really necessary. The small, curved, arrow head is there to remind us that phasors 'rotate' counter-clockwise.

figure 5.3

Step 2:

The current, $\overline{I}_R$, flowing in the resistive branch is *in phase with the supply voltage* and, so, is also drawn along the horizontal positive axis (figure 5.4).

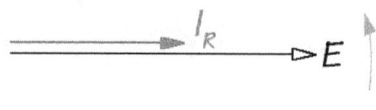

figure 5.4

Step 3:

The current, $\overline{I}_L$, flowing in the inductive branch *lags* the voltage by 90° (remember CI**VIL**) so, as phasors 'rotate' counter-clockwise, it is drawn 90° clockwise from the reference phasor (figure 5.5)

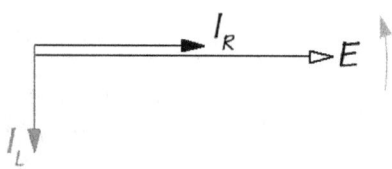

figure 5.5

Step 4:

We know, from Kirchhoff's Current Law that, in a parallel circuit, the supply current ($\overline{I}$) is the sum of the individual branch currents. However, because, in this case, the two currents, $\overline{I}_R$ and $\overline{I}_L$, lie at right-angles to each other, we have to add them *vectorially* (figure 5.6).

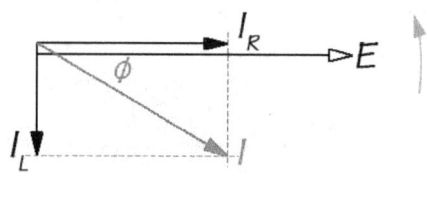

figure 5.6

From the completed phasor diagram, above, we can see that the supply current, $\overline{I}$, is the **phasor-sum** (or **vectorial sum**) of $\overline{I}_R$ and $\overline{I}_L$, which can be found by simply applying Pythagoras's Theorem:

$$\overline{I} = \sqrt{\overline{I}_R^2 + \overline{I}_L^2}$$

To find the phase-angle, we can use the *cosine ratio*:

$$\cos\phi = \frac{adjacent}{hypotenuse} = \frac{\overline{I}_R}{\overline{I}}$$

so...

$$\angle\phi = \cos^{-1}\frac{\overline{I}_R}{\overline{I}}$$

...and, of course, because this is an inductive circuit, it's a **lagging** power factor.

We *could* have used the sine or tangent ratios, too, but by using the cosine ratio, we are able to work out the circuit's **power factor** (which, as we learnt earlier, is the *cosine* of the phase angle) at the same time! That is, we are 'killing two birds with one stone'!

Worked Example 1

The current drawn by the resistive branch of a parallel *R-L* circuit is 15 A, and the current drawn by the inductive branch is 20 A. What is the value of the supply current? Also, what is the circuit's phase-angle?

Solution

Always start by sketching the circuit diagram, and inserting all the values given to you in the question (figure 5.7).

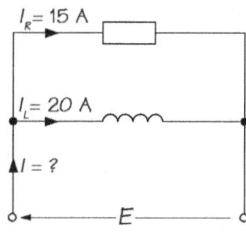

figure 5.7

Next, sketch the phasor diagram, following the steps described above. You *don't* have to draw the phasor diagram to scale (figure 5.8).

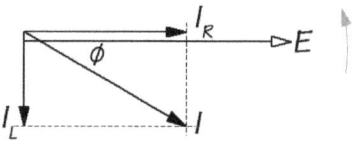

figure 5.8

Now, you can apply Kirchhoff's Current Law, using Pythagoras's Theorem to solve the problem:

$$\overline{I} = \sqrt{\overline{I}_R^2 + \overline{I}_L^2}$$
$$= \sqrt{15^2 + 20^2}$$
$$= \sqrt{625} = 25 \text{ A (Answer)}$$

Since this is an inductive circuit, and the current *lags* the supply-voltage, then the phase-angle will be *lagging*:

$$\angle\phi = \cos^{-1}\frac{\overline{I}_R}{\overline{I}} = \cos^{-1}\frac{15}{25} = \cos^{-1}0.6 = 53.13° \text{ lagging (Answer)}$$

Admittance Diagram

Admittance is the *reciprocal* of impedance, and it is measured in **siemens** (symbol: S). In parallel circuits, the equivalent of an impedance diagram is an **admittance diagram**.

You will recall that, in a **series R-L** diagram, we changed the voltage phasor diagram into an *impedance diagram* simply by dividing throughout by the reference phasor which, in that case, was the supply current, $\overline{I}$.

Well, let's follow *exactly the same rule* and, again: *divide the phasor diagram by the reference phasor*. For our parallel circuit, however, the reference phasor is the supply-voltage, $\overline{E}$, so let's see what happens this time.

Drawing the Admittance Diagram
Step 1:

We start by drawing the circuit's phasor diagram, following the steps already explained (figure 5.9).

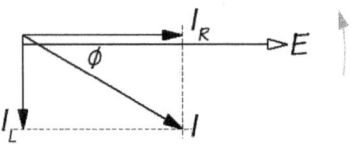

figure 5.9

Step 2:

Next, we *divide each of the current phasors by the reference phasor ($\overline{E}$)* (figure 5.10).

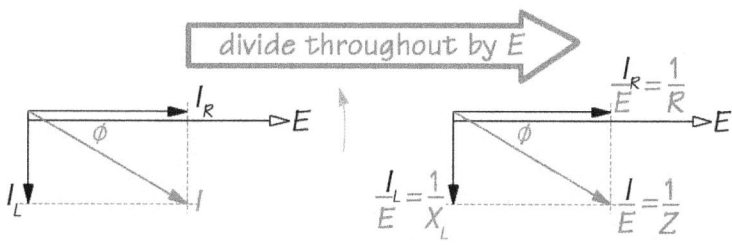

figure 5.10

As you can see, what we now have are expressions for the *reciprocals* of resistance, inductive reactance, and impedance. We call these **conductance**, **inductive susceptance**, and **admittance**, respectively —hence the term, **'admittance diagram'**.

Figure 6.11 shows exactly the same admittance diagram, but expressed directly in terms of **conductance *(G)*, inductive susceptance *(B_L)*,** and **admittance *(Y)*,** each of which is expressed in siemens (S):

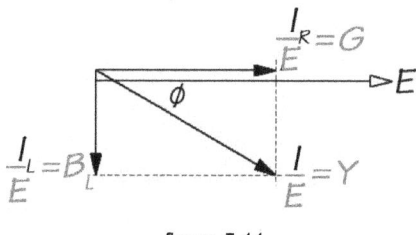

figure 5.11

As you can see, we have ended up with the *reciprocals* of resistance, inductive reactance, and impedance. The resulting diagram is called an **admittance diagram** (sometimes called an *'admittance triangle'*), and is useful because it generates the following equations:

$$\frac{\overline{I_R}}{\overline{E}} = G = \frac{1}{R} \qquad \frac{\overline{I_L}}{\overline{E}} = B_L = \frac{1}{X_L} \qquad \frac{\overline{I}}{\overline{E}} = Y = \frac{1}{Z}$$

If we apply Pythagoras's Theorem to the admittance diagram, we get the following relationship:

$$Y^2 = G^2 + B_L^2$$

...if we wish, we can also express this equation in terms of impedance, resistance, and inductive reactance, as follows::

$$\left(\frac{1}{Z}\right)^2 = \left(\frac{1}{R}\right)^2 + \left(\frac{1}{X_L}\right)^2$$

> ...which is *exactly* equivalent to the following equation that we are already familiar with for calculating the total resistance of a parallel d.c. circuit:
> $$\frac{1}{R} = \frac{1}{R_1} + \frac{1}{R_2}$$

Manipulating the 'admittance' equation, then gives us:

$$\boxed{\frac{1}{Z} = \sqrt{\frac{1}{R^2} + \frac{1}{X_L^2}}}$$

...from which, if we invert the answer, we can find the **impedance** of the parallel circuit.

We can also find the circuit's **phase-angle**, using basic trigonometry, utilising either the *sine*, *cosine*, or *tangent* ratios. In practice, for a reason we'll see shortly, the best choice is always to use the *cosine*:

$$\cos\phi = \frac{\text{adjacent}}{\text{hypotenuse}} = \frac{\left(\frac{1}{R}\right)}{\left(\frac{1}{Z}\right)} = \frac{Z}{R}$$

$$\boxed{\angle\phi = \cos^{-1}\frac{Z}{R}}$$

You will notice that this is rather different from the corresponding equation for a series *R-L* circuit! This is because the effective impedance is *smaller* than either the resistance or the inductive reactance (in just the same way that the effective resistance of a d.c. parallel circuit is always less than either of the branch resistances).

> **Important!**
> Dividing a **current phasor diagram** by voltage produces an **admittance diagram** which generates each of the equations shown above. **So you don't have to learn *any* of these equations** —they can all be generated *provided you learn how to draw the phasor and impedance diagrams!*

However, *if you prefer*, you can rewrite the above equations directly in terms conductance, inductive susceptance, and conductance, as follows:

$$\boxed{\dfrac{\bar{I}_R}{\bar{E}} = G} \qquad \boxed{\dfrac{\bar{I}_L}{\bar{E}} = B_L} \qquad \boxed{\dfrac{\bar{I}}{\bar{E}} = Y}$$

By applying Pythagoras's Theorem, you can also find the following relationship:

$$Y^2 = G^2 + B_L^2$$

$$\boxed{Y = \sqrt{G^2 + B_C^2}}$$

We can also find the circuit's **phase-angle**, using basic trigonometry, utilising either the *sine*, *cosine*, or *tangent* ratios. As usual, the best choice is always to use the *cosine*, because it also tells you what the circuit's power factor happens to be, should you need to know:

$$\cos\phi = \dfrac{\text{adjacent}}{\text{hypotenuse}} = \dfrac{G}{Y}$$

$$\boxed{\angle\phi = \cos^{-1}\dfrac{G}{Y}}$$

> In many respects, it's far more convenient to work in terms of conductance, inductive susceptance, and admittance, as it avoids the complications of having to work with reciprocals (fractions). But it's entirely up to you; you can do whichever you are more comfortable with.

Worked Example 2

A parallel circuit of resistance 5 Ω in parallel with an inductance 0.02 H, is connected across a 230-V, 50 Hz, a.c. supply. Calculate each of the following:

  a. inductive reactance
  b. impedance
  c. current through the resistive branch
  d. current through the inductive branch
  e. supply current
  f. phase angle of the circuit

Solution

As always, the first set in solving *any* circuit problem is to sketch the circuit diagram, and label it with all values supplied in the problem (figure 5.12).

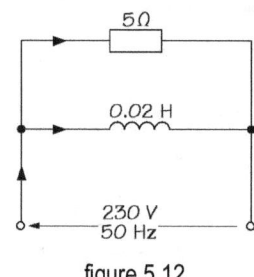

figure 5.12

The next step is to sketch the current phasor diagram, following the steps described earlier (figure 5.13).

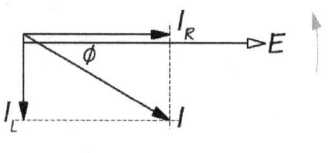

figure 5.13

Next, we must convert the current phasor diagram into an admittance diagram (figure 5.14) by dividing throughout by the reference quantity —i.e. the supply voltage. This generates all the equations that we need to solve the problem (without you having to remember them!):

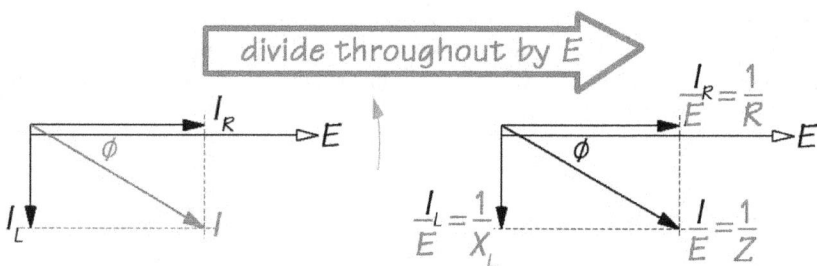

figure 5.14

a. To find the **inductive-reactance** (X_L) of the circuit, we start by looking at the equations generated when we constructed the admittance diagram. There's only one equation with inductive reactance shown, $\dfrac{\overline{I_L}}{\overline{E}} = \dfrac{1}{X_L}$. Unfortunately, we don't know the value of $\overline{I_L}$, so we can't use this formula. So we must fall back on the basic equation for inductive reactance, as follows:

$$X_L = 2\pi f L = 2\pi \times 50 \times 0.02 = 6.28 \ \Omega \ \text{(Answer a.)}$$

b. To find the impedance, we can now use the equation generated by the admittance diagram:

$$\frac{1}{Z} = \sqrt{\frac{1}{R^2} + \frac{1}{X_L^2}}$$

$$\frac{1}{Z} = \sqrt{\frac{1}{5^2} + \frac{1}{6.28^2}} = \sqrt{\frac{1}{25} + \frac{1}{39.44}} = \sqrt{\frac{39.44 + 25}{25 \times 39.44}}$$

$$\frac{1}{Z} = \sqrt{\frac{64.44}{986}} = \sqrt{0.0654} = 0.2557$$

$$Z = \frac{1}{0.2557} = 3.91\,\Omega \text{ (Answer b.)}$$

This answer makes sense, because, in a parallel circuit, the impedance must be lower than either the resistance or the reactance (in exactly the same way that the equivalent resistance of a d.c. parallel circuit must be less than the resistance of any branch).

The alternative, and somewhat simpler, way of solving this part of the question, is work directly with *conductance, inductive susceptance*, and *admittance*.

First, find the **conductance** and the **inductive susceptance**:

$$G = \frac{1}{R} = \frac{1}{5} = 0.20\,\text{S} \quad \text{and} \quad B_L = \frac{1}{X_L} = \frac{1}{6.28} = 0.16\,\text{S}$$

Next, apply the Pythagoras's Theorem equation for **admittance**:

$$Y = \sqrt{G^2 + B_L^2} = \sqrt{0.2^2 + 0.16^2} = \sqrt{0.0656} = 0.256\,\text{S}$$

Finally, the **impedance** is the inverse of the admittance:

$$Z = \frac{1}{Y} = \frac{1}{0.256} = 3.91\,\Omega \text{ (Answer b.)}$$

---

c. To find the current through the resistive branch, we use the following equation generated by the admittance diagram:

$$\frac{1}{R} = \frac{\overline{I}_R}{\overline{E}}$$

$$\text{rearranging, } \overline{I}_R = \frac{\overline{E}}{R} = \frac{230}{5} = 46\,\text{A (Answer c.)}$$

> Again, if you prefer, you can use *conductance*, rather than resistance, to work out the current through the resistive branch:
>
> $$\frac{\overline{I}_R}{\overline{E}} = G$$
>
> rearranging, $\overline{I}_R = \overline{E}\,G = 230 \times 0.2 = 46$ A (Answer c.)

d. Again, using the equation generated by the admittance diagram:

$$\frac{1}{X_L} = \frac{\overline{I}_L}{\overline{E}}$$

rearranging, $\overline{I}_L = \dfrac{\overline{E}}{X_L} = \dfrac{230}{6.28} = 36.62$ A (Answer d.)

> Again, if you prefer, you can use *conductance*, rather than resistance, to work our the current through the resistive branch:
>
> $$\frac{\overline{I}_L}{\overline{E}} = B_L$$
>
> rearranging, $\overline{I}_L = \overline{E}\,B_L = 230 \times 0.16 = 36.8$ A (Answer d.)
>
> (*minor difference in answer due to rounding up/down*)

e. To find the supply current, we can apply Pythagoras' Theorem to the current phasor diagram,

$$\overline{I} = \sqrt{\overline{I}_R^2 + \overline{I}_L^2} = \sqrt{46^2 + 36.62^2} = \sqrt{2116 + 1341} = 58.78 \text{ A (Answer e.)}$$

f. Using the cosine function:

$$\angle\phi = \cos^{-1}\frac{Z}{R} = \cos^{-1}\frac{3.911}{5} = \cos^{-1} 0.7822 = 38.54° \text{ lagging (Answer f.)}$$

(*'Lagging'*, because the *supply current lags the supply voltage* in an inductive circuit.)

> Once again, if you prefer to work in terms of *admittance* and *conductance*, then:
>
> $$\angle \phi = \cos^{-1} \frac{G}{Y} = \cos^{-1} \frac{0.2}{0.256} = \cos^{-1} 0.7813$$
> $$= 38.62° \text{ lagging (Answer f.)}$$
>
> *(minor difference in answer, due to rounding up/down)*

Power Diagram

To create a **power diagram** (or '**power triangle**') for a parallel *R-L* circuit we simply *multiply* throughout by the reference phasor (figure 5.15).

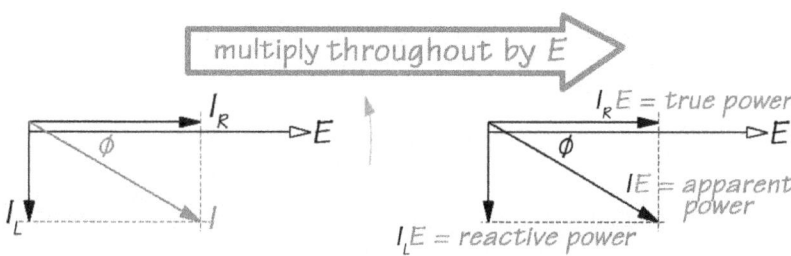

figure 5.15

This action generates the *exactly* same power equations as we have already seen for a *series R-L* circuit. As always, if you can draw the circuit's current phasor diagram, and convert it into a power diagram, *then you can **derive** all the following equations, without having to remember them:*

$$\boxed{\overline{I}_R \overline{E} = \text{true power}} \quad \boxed{\overline{I}_L \overline{E} = \text{reactive power}} \quad \boxed{\overline{I}\,\overline{E} = \text{apparent power}}$$

By applying Pythagoras's Theorem, we can obtain an equation for apparent power (in volt amperes) in terms of true power (in watts) and reactive power (in reactive volt amperes):

$$\boxed{(\text{apparent power})^2 = (\text{true power})^2 + (\text{reactive power})^2}$$

Finally, to find the **power factor** of the circuit, we simply need to find the cosine of the phase angle, which is:

$$\cos \phi = \frac{\text{adjacent}}{\text{hypotenuse}} = \frac{\text{true power}}{\text{apparent power}}$$

Parallel R-C circuits

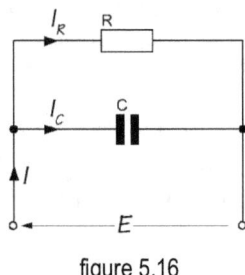

figure 5.16

We know that in a purely resistive circuit, the current and voltage are in phase with each other; and, in a purely capacitive circuit, the current *leads* the voltage by 90°. So, in a *parallel R-C* circuit (figure 5.26), the *current must lead the voltage by some angle between 0° and 90°* —this angle is called the circuit's **phase-angle** (symbol: ϕ, pronounced *'phi'*).

When a potential difference ($\overline{E}$) flows through a **parallel R-C circuit**, a current, $\overline{I}_R$, will flow through the resistive branch, and a current, $\overline{I}_C$, will flow through the capacitive branch.

We needn't go through the step-by-step process of creating the current phasor diagram for this circuit, as it is practically the same as the procedure we have already gone through to create the phasor diagram for an *R-L* circuit. The important difference, of course, is that, for an *R-C* circuit, the current through the capacitive branch, $\overline{I}_C$, *leads* the supply voltage.

So, the finished current phasor diagram for a parallel *R-C* circuit will look like figure 5.17.

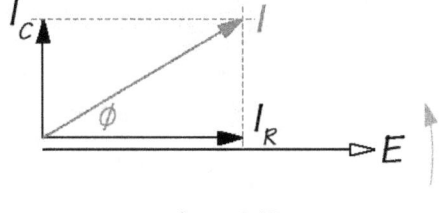

figure 5.17

From the completed phasor diagram, above, we can see that the supply current, $\overline{I}$, is the **phasor-sum** (or **vectorial sum**) of $\overline{I}_R$ and $\overline{I}_C$, which can be found using Pythagoras's Theorem:

$$\overline{I} = \sqrt{\overline{I}_R^2 + \overline{I}_C^2}$$

To find the phase-angle, we can use the *cosine ratio*:

$$\cos\phi = \frac{adjacent}{hypotenuse} = \frac{\overline{I}_R}{\overline{I}}$$

so

$$\angle\phi = \cos^{-1}\frac{\overline{I}_R}{\overline{I}}$$

and, of course, because this is an inductive circuit, it's a **leading** power factor.

Admittance Diagram

Again, by dividing *the phasor diagram by the reference phasor*, $\overline{E}$, we can change the phasor diagram into an admittance diagram (figure 5.18).

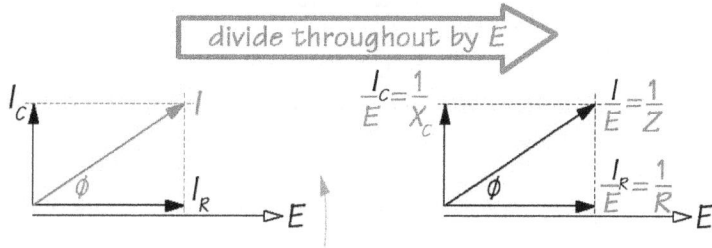

Remember, 'admittance' is the *reciprocal* of impedance:

figure 5.18

The same admittance diagram, expressed directly in terms of **conductance** *(G)*, **capacitive susceptance** *(B_C)*, and **admittance** *(Y)* is shown in figure 5.19.

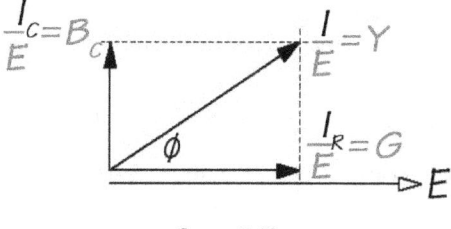

figure 5.19

The resulting diagram is an **admittance diagram** (or *'admittance triangle'*), and is useful because it generates the following equations:

$$\frac{\overline{I}_R}{\overline{E}} = G = \frac{1}{R}$$

$$\frac{\overline{I}_C}{\overline{E}} = B_c = \frac{1}{X_C}$$

$$\frac{\overline{I}}{\overline{E}} = Y = \frac{1}{Z}$$

Also from the admittance diagram, you can also see that the admittance can be calculated by applying Pythagoras's Theorem:

$$\frac{1}{Z} = \sqrt{\left(\frac{1}{R}\right)^2 + \left(\frac{1}{X_C}\right)^2}$$

We can also find the circuit's **phase-angle**, using basic trigonometry, utilising the *cosine* ratio.

$$\cos\phi = \frac{\text{adjacent}}{\text{hypotenuse}} = \frac{\left(\frac{1}{R}\right)}{\left(\frac{1}{Z}\right)} = \frac{Z}{R}$$

$$\angle\phi = \cos^{-1}\frac{Z}{R}$$

Important!

Dividing a **current phasor diagram** by voltage produces an **admittance diagram** which generates each of the equations shown above. **So you don't have to learn *any* of these equations** —they can all be generated *provided you learn how to draw the phasor and admittance diagrams!*

Again, you may prefer to work in terms of conductance, capacitive susceptance, and admittance. In which case,

$$\frac{\overline{I}_R}{\overline{E}} = G$$

$$\frac{\overline{I}_C}{\overline{E}} = B_C$$

$$\frac{\overline{I}}{\overline{E}} = Y$$

Also from the admittance diagram, you can also see that the admittance can be calculated by applying Pythagoras's Theorem:

$$\boxed{Y = \sqrt{G^2 + B_C{}^2}}$$

We can also find the circuit's **phase-angle**, using basic trigonometry, utilising the *cosine* ratio.

$$\cos\phi = \frac{\text{adjacent}}{\text{hypotenuse}} = \frac{G}{Y}$$

$$\boxed{\angle\phi = \cos^{-1}\frac{G}{Y}}$$

Once again, in many respects, it's more convenient to work in terms of conductance, inductive susceptance, and admittance, as it avoids the complications of having to work with fractions. But it's entirely up to you!

Power Diagram

To create a **power diagram** (or '**power triangle**') for a parallel *R-C* circuit, we simply multiply throughout by the reference phasor (figure 5.20).

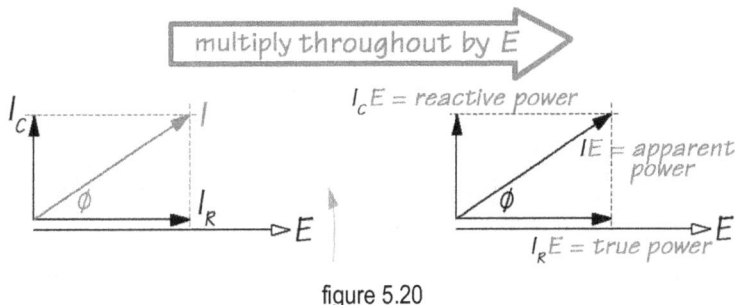

figure 5.20

This action generates the same power equations as we have already seen for a *series R-L* circuit. As always, if you can draw the circuit's current phasor diagram, and convert it into a power diagram, *then you can **derive** all the following equations, without having to remember them:*

$$\boxed{\overline{I}_R \overline{E} = \text{true power}} \quad \boxed{\overline{I}_C \overline{E} = \text{reactive power}} \quad \boxed{\overline{I}\,\overline{E} = \text{apparent power}}$$

By applying Pythagoras's Theorem, we can obtain an equation for apparent power (in volt amperes) in terms of true power (in watts) and reactive power (in reactive volt amperes):

$$\boxed{(\text{apparent power})^2 = (\text{true power})^2 + (\text{reactive power})^2}$$

Finally, to find the **power factor** of the circuit, we simply need to find the cosine of the phase angle, which is:

$$\cos\phi = \frac{\text{adjacent}}{\text{hypotenuse}} = \frac{\text{true power}}{\text{apparent power}}$$

...and, of course, this time it's a **leading** power factor.

Worked Example 3

A parallel *R-C* circuit comprises a 30-Ω resistor in parallel with a capacitor of capacitive-reactance 20 Ω, connected across a 120-V supply. Calculate the circuit's (a) true power, (b) reactive power, (c) apparent power, and (d) power factor.

Solution

As always, the first step is to sketch a circuit diagram, with all the supplied information inserted (figure 5.21).

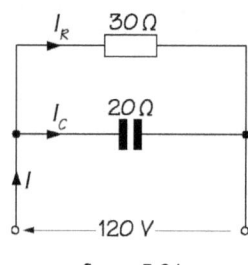

figure 5.21

Before we can work out the true-, reactive-, and apparent power, *we need to work out the current flowing in each branch.* So we must sketch the current phasor diagram for the circuit, and divide throughout by the reference phasor *($\overline{E}$)* to produce the **admittance diagram** (figure 5.22). *All* the equations you will need are then generated without you having to remember them:

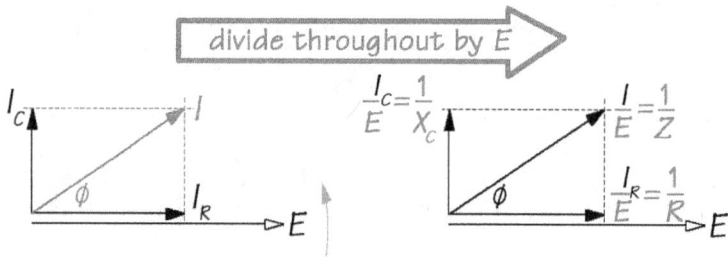

figure 5.22

To find the current through the resistor, $\overline{I}_R$:

$$\frac{\overline{I}_R}{\overline{E}} = \frac{1}{R} \quad \text{rearranging:} \quad \overline{I}_R = \frac{\overline{E}}{R} = \frac{120}{30} = 4 \text{ A}$$

To find the current through the capacitor, $\overline{I}_C$:

$$\frac{\overline{I}_C}{\overline{E}} = \frac{1}{X_C} \quad \text{rearranging:} \quad \overline{I}_C = \frac{\overline{E}}{X_C} = \frac{120}{20} = 6 \text{ A}$$

To find the supply current, since we don't know the circuit's impedance, we apply Pythagoras's Theorem:

$$I = \sqrt{I_R^2 + I_C^2} = \sqrt{4^2 + 6^2} = 7.21 \text{ A}$$

Now we can convert the current phasor diagram into a power diagram (figure 5.23)., by multiplying throughout by the reference phasor $(\overline{E})$:

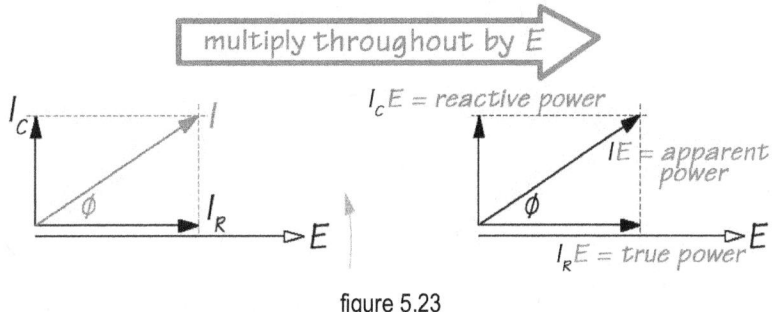

figure 5.23

To find the true power:

$$\text{true power} = \overline{I}_R \overline{E} = 4 \times 120 = 480 \text{ W (Answer a.)}$$

To find the reactive power:

$$\text{reactive power} = \overline{I}_C \overline{E} = 6 \times 120 = 720 \text{ var (Answer b.)}$$

To find the apparent power:

$$\text{apparent power} = \overline{IE} = 7.21 \times 120 \square 865 \text{ V} \cdot \text{A (Answer c.)}$$

An alternative method, would be to apply Pythagoras's Theorem, as follows:

$$\text{apparent power} = \sqrt{\text{true power}^2 + \text{reactive power}^2}$$
$$= \sqrt{480^2 + 720^2} = 865 \text{ V} \cdot \text{A}$$

Finally, to find the power factor, we use the cosine ratio:

$$\text{power factor} = \cos\phi = \frac{\text{adjacent}}{\text{hypotenuse}} = \frac{\text{true power}}{\text{apparent power}}$$

$$= \frac{480}{865} = 0.555 \text{ leading (Answer d.)}$$

'Leading' because, for an R-C circuit, the load current leads the supply voltage.

Parallel R-L-C circuits

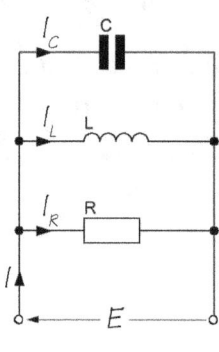

figure 5.24

When a potential difference ($\overline{E}$) is applied across a **parallel R-L-C circuit** (figure 5.24), a current, $\overline{I}_R$, will flow through the resistive branch of the circuit, a current, $\overline{I}_L$, will flow through the inductive branch, and a current, $\overline{I}_C$, will flow through the capacitive branch.

Drawing the Phasor Diagram

Step 1:

In a parallel circuit, the **supply voltage** ($\overline{E}$) is common to each branch and, so, this is chosen as the **reference phasor**. As we have learnt, the reference phasor is *always drawn along the horizontal positive axis*, and its also normally drawn fairly long in order to distinguish it from the others (figure 5.25).

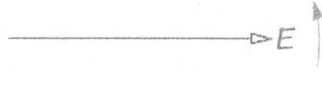

figure 5.25

Step 2:

The current, $\overline{I}_R$, flowing through the resistive branch, is *in phase with the supply voltage* and, so, is also drawn along the horizontal positive axis (figure 5.26).

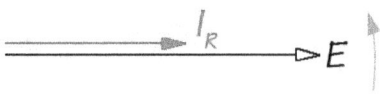

figure 5.26

Step 3:

The current, $\overline{I}_L$, flowing through the inductive branch *lags the supply voltage by 90°* (remember CI**VIL**), so is drawn 90° clockwise from the reference phasor (figure 5.27).

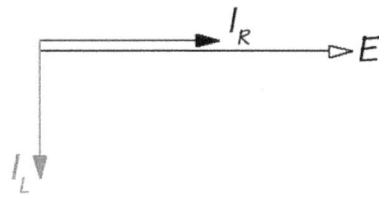

figure 5.27

Step 4:

The current, $\overline{I}_C$, flowing through the capacitive branch *leads the supply voltage by 90°* (remember **CIV**IL), so is drawn 90° counter-clockwise from the reference phasor (figure 5.28). ***Always*** **draw $\overline{I}_C$ shorter than $\overline{I}_L$ (or vice-versa)**, or we will end up with a unique condition called '**resonance**' —just as we did in the *series R-L-C* circuit!

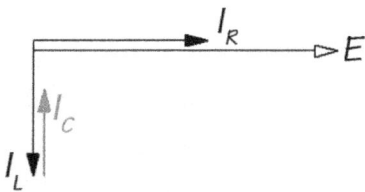

figure 5.28

Step 5:

We know from Kirchhoff's Current Law that, in a parallel circuit, the supply current is the sum of the individual branch currents and, for a.c. circuits, we have to add them *vectorially*.

As always, it's a little more difficult to add *three* phasors. As $\overline{I}_L$ and $\overline{I}_C$ lie in opposite directions, the simplest thing to do is to start by *subtracting them* and, then, vectorially-add the *difference* to phasor $\overline{I}_R$.

The snag is, of course, that we might not know whether $\overline{I}_L$ is bigger than $\overline{I}_C$, or *vice-versa*! Fortunately, *it doesn't matter!* Once again, the purpose of the phasor diagram is to *generate equations*, *not* to accurately represent the *actual* conditions in the circuit to scale! And the phasor diagram will *always* generate the correct equations, whether or not $\overline{I}_L$ is bigger than $\overline{I}_C$!

So, the simplest solution is to get into the habit of *always drawing $\overline{I}_L$ longer than $\overline{I}_C$* —but whatever you do, **never draw them the same length!!**

So, as explained, start by subtracting $\overline{I}_C$ from $\overline{I}_L$, and *then* vectorially add the resultant to $\overline{I}_R$ to give the completed phasor diagram, as shown in figure 5.29.

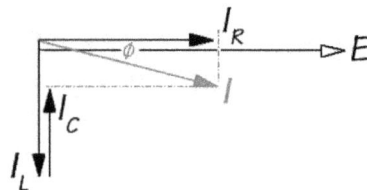

figure 5.29

From the completed phasor diagram, above, we can see that the supply current, $\overline{I}$, is the **phasor-sum** (or **vectorial sum**) of $\overline{I}_R, \overline{I}_L,$ and $\overline{I}_C$, which can then be found using Pythagoras's Theorem:

$$\boxed{\overline{I} = \sqrt{\overline{I}_R^2 + (\overline{I}_L - \overline{I}_C)^2}}$$

To find the phase-angle, we can use the *cosine ratio*:

$$\cos\phi = \frac{adjacent}{hypotenuse} = \frac{\overline{I}_R}{\overline{I}}$$

If $\overline{I}_L$ really is larger than $\overline{I}_C$, then the circuit is predominantly inductive and, so, the resulting phase-angle will be **lagging**. On the other hand, if $\overline{I}_C$ happens to be larger than $\overline{I}_L$, then the circuit will be predominantly capacitive and, so, the resulting phase-angle will be **leading**

Admittance Diagram

How can we now proceed to find out further equations for solving a parallel *R-L-C* circuit?

Again, the answer is by means of an **admittance diagram**:

Drawing the Admittance Diagram
Step 1:

We start by drawing the circuit's phasor diagram, following the steps already explained (figure 5.30).

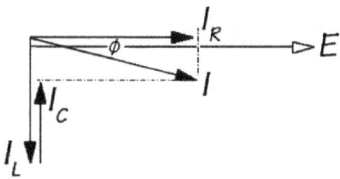

figure 5.30

Step 2:

Next, we *divide each of the voltage phasors by the reference phasor* ($\overline{E}$) — as shown in figure 5.31.

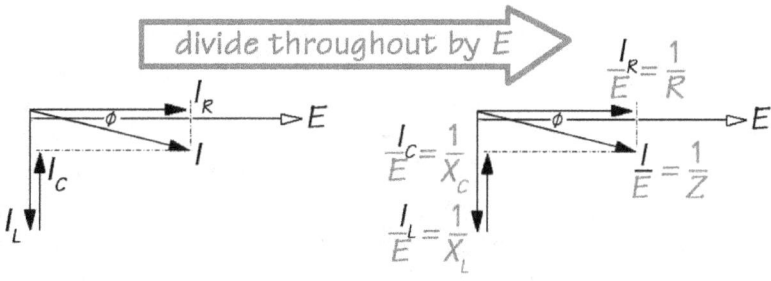

figure 5.31

The resulting diagram is an **admittance diagram** (or *'admittance triangle'*), and is useful because it generates the following equations:

$$\frac{\overline{I}_R}{\overline{E}} = G = \frac{1}{R}$$

$$\frac{\overline{I}_L}{\overline{E}} = B_L = \frac{1}{X_L}$$

$$\frac{\overline{I}_C}{\overline{E}} = B_C = \frac{1}{X_C}$$

$$\frac{\overline{I}}{\overline{E}} = Y = \frac{1}{Z}$$

Also from the admittance diagram, you can also see that the admittance is also the *vectorial sum of conductance ($1/R$) and capacitive susceptance ($1/X_C$)*, which can be calculated by applying Pythagoras' Theorem:

$$\left(\frac{1}{Z}\right)^2 = \left(\frac{1}{R}\right)^2 + \left(\frac{1}{X_L} - \frac{1}{X_C}\right)^2$$

$$\frac{1}{Z} = \sqrt{\left(\frac{1}{R}\right)^2 + \left(\frac{1}{X_L} - \frac{1}{X_C}\right)^2}$$

We can also find the circuit's **phase-angle**, using basic trigonometry, utilising the *cosine* ratio:

$$\cos\phi = \frac{\text{adjacent}}{\text{hypotenuse}} = \frac{\left(\frac{1}{R}\right)}{\left(\frac{1}{Z}\right)}$$

$$\angle\phi = \cos^{-1}\frac{Z}{R}$$

Important!
Dividing a **current phasor diagram** by voltage produces an **admittance diagram** which generates each of the equations shown above. **So you don't have to learn *any* of these equations** —they can all be generated *provided you learn how to draw the phasor and impedance diagrams!*

Once again, you may prefer to work in terms of conductance, inductive susceptance, capacitive susceptance, and admittance (after all, it is easier!), in which case the following equations apply (figure 5.32).

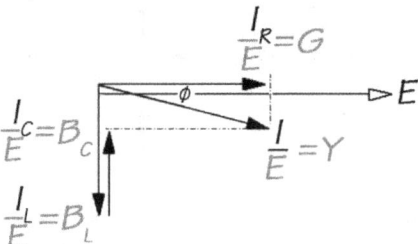

figure 5.32

$$\frac{\overline{I_R}}{\overline{E}} = G \qquad \frac{\overline{I_L}}{\overline{E}} = B_L \qquad \frac{\overline{I_C}}{\overline{E}} = B_C$$

$$\frac{\overline{I}}{\overline{E}} = Y$$

Also from the admittance diagram, you can also see that the admittance is also the *vector sum of conductance (G), inductive susceptance (B_L),* and *capacitive susceptance (B_C),* which can be calculated by applying Pythagoras's Theorem:

$$Y^2 = G^2 + (B_L - B_C)^2$$

$$\boxed{Y = \sqrt{G^2 + (B_L - B_C)^2}}$$

We can also find the circuit's **phase-angle**, using basic trigonometry, utilising the *cosine* ratio:

$$\cos\phi = \frac{\text{adjacent}}{\text{hypotenuse}} = \frac{G}{Y}$$

$$\boxed{\angle\phi = \cos^{-1}\frac{G}{Y}}$$

Power Diagram

To create a **power diagram** (or '**power triangle**') for a parallel *R-C* circuit, we simply multiply throughout by the reference phasor (figure 5.33).

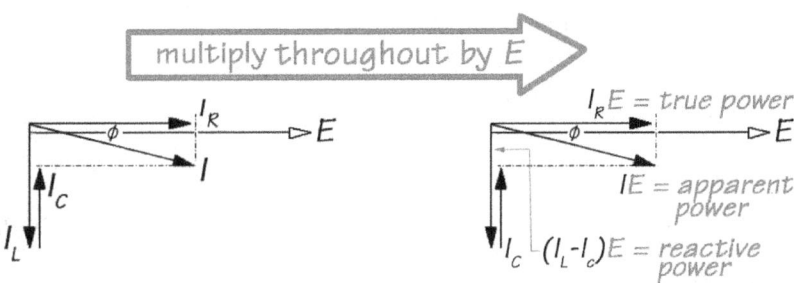

figure 5.33

This action generates the same power equations as we have already seen for a *series R-L-C* circuit. As always, if you can draw the circuit's current phasor diagram, and convert it into a power diagram, *then you can **derive** all the following equations, without having to remember them:*

$$\boxed{\overline{I}_R \overline{E} = \text{true power}} \qquad \boxed{\overline{I}\,\overline{E} = \text{apparent power}}$$

$$\boxed{(\overline{I}_L - \overline{I}_C)\overline{E} = \text{reactive power}}$$

Note, that for **reactive power**, we have had to multiply the supply voltage by the *difference* between the inductive and capacitive currents.

By applying Pythagoras's Theorem, we can obtain an equation for apparent power (in volt amperes) in terms of true power (in watts) and reactive power (in reactive volt amperes):

$$(\text{apparent power})^2 = (\text{true power})^2 + (\text{reactive power})^2$$

Finally, to find the **power factor** of the circuit, we simply need to find the cosine of the phase angle, which is:

$$\cos\phi = \frac{\text{adjacent}}{\text{hypotenuse}} = \frac{\text{true power}}{\text{apparent power}}$$

Worked Example 4

A parallel *R-L-C* circuit comprises a 6-Ω purely-resistive branch, a 19-mH purely-inductive branch, and a 66-µF purely-capacitive branch, supplied by a 48-V, 100-Hz, supply. Calculate:

  a. each of the branch currents,
  b. the supply current,
  c. the impedance,
  d. true power of the complete circuit.

Solution

As always, the first step is to sketch the circuit, with all the supplied information inserted (figure 5.34).

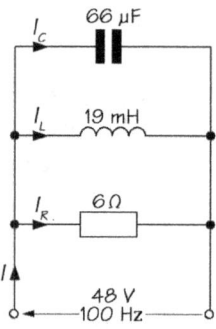

figure 5.34

Before proceeding any further, we should determine the inductive reactance and capacitive reactance of the inductive and capacitive branches, as we will need to use these values:

$$X_L = 2\pi f L = 2 \times \pi \times 100 \times (19 \times 10^{-3}) = 12\,\Omega$$

$$X_C = \frac{1}{2\pi f C} = \frac{1}{2\pi \times 100 \times (66 \times 10^{-6})} = 24\,\Omega$$

The next step is to sketch the current phasor diagram, and convert this into an admittance diagram, by dividing throughout by the reference phasor $(\overline{E})$ (figure 5.35).

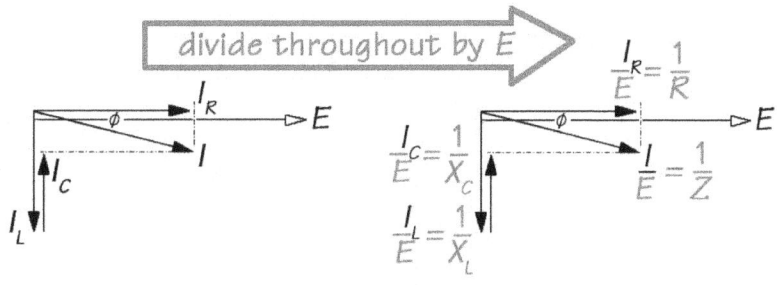

figure 5.35

To find the current through the **resistive** branch:

$$\frac{\overline{I_R}}{\overline{E}} = \frac{1}{R} \quad \text{from which:} \quad \overline{I_R} = \frac{\overline{E}}{R} = \frac{48}{6} = 8\,\text{A (Answer a.)}$$

To find the current through the **inductive** branch:

$$\frac{\overline{I_L}}{\overline{E}} = \frac{1}{X_L} \quad \text{from which:} \quad \overline{I_L} = \frac{\overline{E}}{X_L} = \frac{48}{12} = 4\,\text{A (Answer a.)}$$

To find the current through the **capacitive** branch:

$$\frac{\overline{I_C}}{\overline{E}} = \frac{1}{X_C} \quad \text{from which:} \quad \overline{I_C} = \frac{\overline{E}}{X_C} = \frac{48}{24} = 2\,\text{A (Answer a.)}$$

We now need to refer back to the current phasor diagram, and apply Pythagoras's Theorem, to determine the supply current:

$$I = \sqrt{I_R^2 + (I_L - I_C)^2} = \sqrt{8^2 + (4-2)^2} = \sqrt{68} = 8.25\,\text{A (Answer b.)}$$

To find the impedance of the circuit, we can use the appropriate equation generated by the admittance diagram:

$$\frac{\overline{I}}{\overline{E}} = \frac{1}{Z} \quad \text{from which:} \quad Z = \frac{\overline{E}}{\overline{I}} = \frac{48}{8.25} = 5.82\ \Omega\ \text{(Answer c.)}$$

To find the true power of the circuit, we must convert the current phasor diagram into a power diagram, by multiplying throughout by the reference phasor $(\overline{E})$ —as shown in figure 5.36.

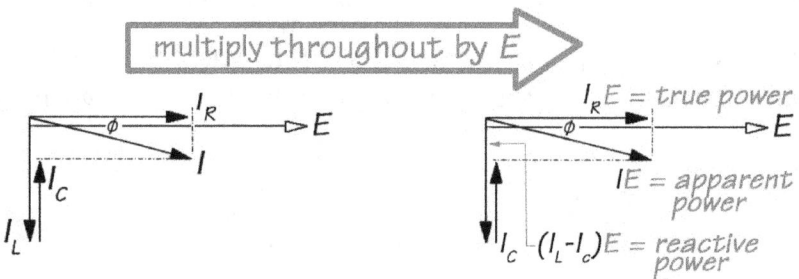

figure 5.36

From the various equations generated, we can use the following:

$$\text{true power} = I_R E = 8 \times 48 = 384\ \text{W (Answer d.)}$$

Review what you have learnt

Now that you have completed this chapter, go back and look at the **objectives** listed at the beginning. If you place a question mark at the end of each objective, and ask yourself, *'Can I...'*, then those objectives become **test items**. If you can answer each test item correctly, then you have successfully completed this book.

Annexe 2

Summary of equations for parallel A.C. circuits

During chapter 6, we learnt how to construct a phasor diagram for a **parallel a.c. circuit**. Once we have learnt to construct a phasor diagram for a **parallel a.c. circuit**, solving an 'electrical' problem becomes a simple matter of applying either *basic geometry* (Pythagoras's Theorem) or *basic trigonometry* (sine, cosine, or tangent ratios), where the phasor diagrams are simple, right-angled, triangles!

Applying **Pythagoras's Theorem** to each of these current phasor diagrams, will give us the following relationships:

Parallel *R-L* circuit:	Parallel *R-C* circuit:	Parallel *R-L-C* circuit:
figure 29.13	figure 29.14	figure 29.15
$\overline{I}^2 = \overline{I}_R^2 + \overline{I}_L^2$	$\overline{I}^2 = \overline{I}_R^2 + \overline{I}_C^2$	$\overline{I}^2 = \overline{I}_R^2 + (\overline{I}_L - \overline{I}_C)^2$

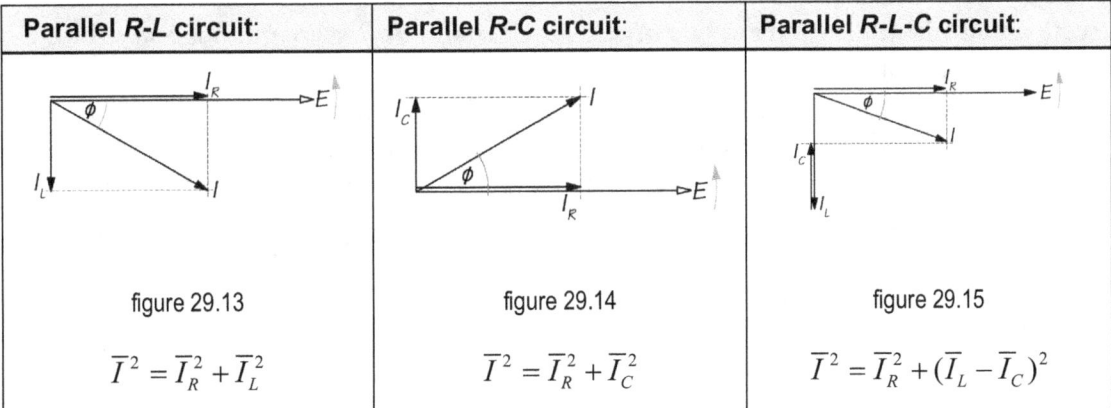

From which we can derive the following equations:

$\overline{I} = \sqrt{\overline{I}_R^2 + \overline{I}_L^2}$	$\overline{I} = \sqrt{\overline{I}_R^2 + \overline{I}_C^2}$	$\overline{I} = \sqrt{\overline{I}_R^2 + (\overline{I}_L - \overline{I}_C)^2}$
$\overline{I}_R = \sqrt{\overline{I}^2 - \overline{I}_L^2}$	$\overline{I}_R = \sqrt{\overline{I}^2 - \overline{I}_C^2}$	$\overline{I}_R = \sqrt{\overline{I} - (\overline{I}_L - \overline{I}_C)^2}$
$\overline{I}_L = \sqrt{\overline{I}^2 - \overline{I}_R^2}$	$\overline{I}_C = \sqrt{\overline{I}^2 - \overline{I}_R^2}$	$(\overline{I}_L - \overline{I}_C) = \sqrt{\overline{I} - \overline{I}_R^2}$

We can also determine the phase angle for each circuit, as follows:

$\angle \phi = \cos^{-1}\left(\dfrac{\overline{I}_R}{\overline{I}}\right)$	$\angle \phi = \cos^{-1}\left(\dfrac{\overline{I}_R}{\overline{I}}\right)$	$\angle \phi = \cos^{-1}\left(\dfrac{\overline{I}_R}{\overline{I}}\right)$

Admittance Diagrams

You will recall that, for a **parallel d.c. circuit**, the total resistance is found as follows:

$$\frac{1}{R} = \frac{1}{R_1} + \frac{1}{R_2} + \frac{1}{R3} + \text{etc.}$$

The *reciprocal* of resistance is termed '**conductance**' (symbol: G), expressed in **siemens** (symbol: **S**). So, we could rewrite the above equation in terms of conductances, that is:

$$G = G_1 + G_2 + G_3 + \text{etc.}$$

In a.c. circuits, the reciprocal of impedance is called '**admittance**' (symbol: Y), and the reciprocal of reactance is called '**susceptance**' (symbol: B). Specifically, the reciprocal of inductive reactance is called '**inductive susceptance**' (symbol: B_L) and the reciprocal of capacitive reactance is called '**capacitive susceptance**' (symbol: B_C).

Generally, in a.c. parallel circuits, it's usually *much* easier to work with admittance, conductance, and susceptance, than it is to work with impedance, resistance, and reactance.

So, to create an **admittance diagram** we *divide the current phasors by the reference phasor* —i.e. by the supply voltage:

Parallel *R-L* circuit:	**Parallel *R-C* circuit:**	**Parallel *R-L-C* circuit:**

From which we get the following equations:

$Y = \dfrac{1}{Z} = \dfrac{\overline{I}}{\overline{E}}$	$Y = \dfrac{1}{Z} = \dfrac{\overline{I}}{\overline{E}}$	$Y = \dfrac{1}{Z} = \dfrac{\overline{I}}{\overline{E}}$
$G = \dfrac{1}{R} = \dfrac{\overline{I}_R}{\overline{E}}$	$G = \dfrac{1}{R} = \dfrac{\overline{I}_R}{\overline{E}}$	$G = \dfrac{1}{R} = \dfrac{\overline{I}_R}{\overline{E}}$
$B_L = \dfrac{1}{X_L} = \dfrac{\overline{I}_L}{\overline{E}}$	$B_C = \dfrac{1}{X_C} = \dfrac{\overline{I}_C}{\overline{E}}$	$B = \dfrac{1}{X} = \dfrac{(\overline{I}_L - \overline{I}_C)}{\overline{E}}$

Applying **Pythagoras's Theorem** to each impedance diagram:

$Y^2 = G^2 + B_L^2$	$Y^2 = G^2 + B_C^2$	$Y^2 = G^2 + (B_L - B_C)^2$
or, if you prefer:	or, if you prefer:	or, if you prefer:

$$\left(\frac{1}{Z}\right)^2 = \left(\frac{1}{R}\right)^2 + \left(\frac{1}{X_L}\right)^2 \quad\quad \left(\frac{1}{Z}\right)^2 = \left(\frac{1}{R}\right)^2 + \left(\frac{1}{X_C}\right)^2 \quad\quad \left(\frac{1}{Z}\right)^2 = \left(\frac{1}{R}\right)^2 + \left(\frac{1}{(X_L - X_C)}\right)^2$$

From which we can find the values of Y, G, B_L, and B_C:

$$Y = \sqrt{G^2 + B_L^2} \quad\quad Y = \sqrt{G^2 + B_C^2} \quad\quad Y = \sqrt{G^2 + (B_L - B_C)^2}$$

$$G = \sqrt{Y^2 - B_L^2} \quad\quad G = \sqrt{Y^2 - B_C^2} \quad\quad G = \sqrt{Y^2 + (B_L - B_C)^2}$$

$$B_L = \sqrt{Y^2 - G^2} \quad\quad B_C = \sqrt{Y^2 - G^2} \quad\quad (B_L - B_C) = \sqrt{Y^2 - G^2}$$

If you prefer, you could *still* work in terms of impedance, resistance, and reactance:

$$\frac{1}{Z} = \sqrt{\left(\frac{1}{R}\right)^2 + \left(\frac{1}{X_L}\right)^2} \quad\quad \frac{1}{R} = \sqrt{\left(\frac{1}{Z}\right)^2 - \left(\frac{1}{X_L}\right)^2} \quad\quad \frac{1}{Z} = \sqrt{\left(\frac{1}{R}\right)^2 + \left(\frac{1}{X_L - X_C}\right)^2}$$

$$\frac{1}{R} = \sqrt{\left(\frac{1}{Z}\right)^2 - \left(\frac{1}{X_L}\right)^2} \quad\quad \frac{1}{R} = \sqrt{\left(\frac{1}{Z}\right)^2 - \left(\frac{1}{X_C}\right)^2} \quad\quad \frac{1}{R} = \sqrt{\left(\frac{1}{Z}\right)^2 - \left(\frac{1}{X_L - X_C}\right)^2}$$

$$\frac{1}{X_L} = \sqrt{\left(\frac{1}{Z}\right)^2 - \left(\frac{1}{R}\right)^2} \quad\quad \frac{1}{X_C} = \sqrt{\left(\frac{1}{Z}\right)^2 - \left(\frac{1}{R}\right)^2} \quad\quad \left(\frac{1}{X_L - X_C}\right) = \sqrt{\left(\frac{1}{Z}\right)^2 - \left(\frac{1}{R}\right)^2}$$

We can also determine the phase angle for each circuit, as follows:

$$\angle\phi = \cos^{-1}\left(\frac{G}{Y}\right) \quad\quad \angle\phi = \cos^{-1}\left(\frac{G}{Y}\right) \quad\quad \angle\phi = \cos^{-1}\left(\frac{G}{Y}\right)$$

Or again, if you prefer: in terms of impedance and resistance:

$$\angle\phi = \cos^{-1}\left(\frac{(1/R)}{(1/Z)}\right) = \frac{Z}{R} \quad\quad \angle\phi = \cos^{-1}\left(\frac{(1/R)}{(1/Z)}\right) = \frac{Z}{R} \quad\quad \angle\phi = \cos^{-1}\left(\frac{(1/R)}{(1/Z)}\right) = \frac{Z}{R}$$

Power Diagrams

We can convert any of the above *current* phasor diagrams into an **power diagram**, by *multiplying throughout by the reference phasor,* $\overline{E}$, as follows. Remember, **'apparent power'** *(S)* is expressed in volt amperes, **'true power'** *(P)* in watts, and reactive power *(Q)* in reactive volt amperes.

Parallel *R-L* circuit:	Parallel *R-C* circuit:	Parallel *R-L-C* circuit:

From which we get the following equations:

$S = \overline{E}\overline{I}$	$S = \overline{E}\overline{I}$	$S = \overline{E}\overline{I}$
$P = \overline{E}\,\overline{I}_R$	$P = \overline{E}\,\overline{I}_R$	$P = \overline{E}\,\overline{I}_R$
$Q = \overline{E}\,\overline{I}_L$	$Q = \overline{E}\,\overline{I}_C$	$Q = \overline{E}(\overline{I}_L - \overline{I}_C)$

Applying **Pythagoras's Theorem** to each power diagram:

$S^2 = P^2 + Q^2$	$S^2 = P^2 + Q^2$	$S^2 = P^2 + Q^2$

From which, we can find the values of apparent power *(S)*, true power *(P)*, and reactive power *(Q)*:

$S = \sqrt{P^2 + Q^2}$	$S = \sqrt{P^2 + Q^2}$	$S = \sqrt{P^2 + Q^2}$
$P = \sqrt{S^2 - Q^2}$	$P = \sqrt{S^2 - Q^2}$	$P = \sqrt{S^2 - Q^2}$
$Q = \sqrt{S^2 - P^2}$	$Q = \sqrt{S^2 - P^2}$	$Q = \sqrt{S^2 - P^2}$

Hopefully, what should have been made abundantly clear from the above is that *every single equation* is derived by applying Pythagoras's Theorem, or the sine, cosine, or tangent ratios to phasor diagrams *which are nothing more than right-angled triangles!*

Accordingly, *there is absolutely no need whatsoever to commit any of these equations to memory!*

Where to go from here...?

If you have completed each of the chapters in this book, then you will have a solid understanding of, and be able to solve, **single-phase series** and **parallel a.c. circuits**.

No doubt, you will now want to move on in order to apply your understanding to the behaviour of more-complicated circuits, i.e. **single-phase series-parallel a.c. circuits**, or **balanced** and **unbalanced three-phase a.c. circuits**?

Unfortunately, drawing a phasor diagram for any, other than the most basic series-parallel circuit can be *very* complicated. So we need to learn alternative techniques for solving series-parallel circuits.

Symbolic Notation

To move beyond the types of circuits described in this book, we need to understand '**symbolic notation**'. This is a purely-mathematical approach which avoids the need to construct their potentially very-complicated phasor diagrams, and involves the application of what is known as 'complex numbers', as represented by the '**j-operator**' (for single-phase circuits) and the '**a-operator**' (for three-phase circuits).

'**Complex numbers**' is a rather unfortunate term because, in fact, 'complex numbers' are *not* in the least 'complicated'! 'Complex numbers' is simply a technical term for a combination of '**real**' and '**imaginary**' numbers. Now, this might *sound* complicated, but they are *very easy to understand*. In this context, 'real' numbers simply represent values measured along the positive and negative horizontal (i.e. the 'real') axis, whereas 'imaginary' numbers simply represent values measured along the positive and negative vertical (i.e. the 'imaginary') axis.

If you have ever used a map, then you will have already used 'real' and 'imaginary' numbers without realizing so, in order to define a point on that map. Map co-ordinates are *always* measured horizontally ('real') and vertically ('imaginary'), in that order (called 'eastings and northings' in map reading). In a similar way, we can specify the location of the arrow-head of any phasor in terms of its 'real' and 'imaginary' co-ordinates. We can then use these 'complex numbers' to add, subtract, multiply, or divide these phasors without the need to use geometry or trigonometry.

The rules for adding, subtracting, multiplying, and dividing 'complex numbers' are also very easy, and can be learnt very quickly. And they can be applied, easily, to any a.c. circuit, no matter how simple or complicated that circuit happens to be.

So, solving a.c. single- and three-phase circuits is rendered very easy through the application of 'symbolic notation'. This topic is covered in depth in my book, *'Electrical Science for Technicians'*, which is readily available from Amazon and other good booksellers.

Other Books by the Same Author

This monograph, *'Alternating Current Explained Simply'*, is based on the content of *'An Introduction to Electrical Science (2nd Edition)'*.

'An Introduction to Electrical Science' is aimed at those embarking upon a electrical career in the electrotechnology industry, particularly electrical apprentices. The book is also ideal for electronics hobbyists. It is written in the same conversational style as this book, and each subject is set it its historical context in order to make it interesting as well as informative.

The book starts with the assumption that the reader has absolutely no knowledge whatsoever of electricity, and each chapter builds of the knowledge gained in the preceding chapters. These chapters include an introduction to the *SI system*, the *electron theory, electric current, potential and potential difference, resistance, energy and power, Ohm's Law*, and so on. The book also includes chapters on *measuring instruments, cells and batteries, magnetism, electromagnetism*, etc.

A more-advanced book, *'Electrical Science for Technicians'* continues on from where the first book leaves off, and covers topics in greater depth. It includes chapters on *network theorems*, the application of *symbolic notation* (the j- and a-operators) *to single-phase series-parallel a.c. circuits and to unbalanced three-phase a.c. systems, transformer theory, d.c. and a.c. motors and generators*, and *lighting*. It is written in the same, conversational style as the first book and, again, the subjects are presented within their historical contex

This is what **Amazon** customers have said about the first edition of *'An Introduction to Electrical Science'*:

> *Should be a mandatory textbook for anyone taking an electrical engineering course. If you work through the entire text (and the subsequent textbook: 'Electrical Science for Technicians') you will have a better fundamental understanding of electrical science than most experienced technicians. A really solid introduction that will make your life so much easier if/when you move onto more advanced topics within EEE. The author has clearly taught this subject for a number of years, and knows what elements to concentrate on and to clarify. It sounds like he's berating you a bit at times (he does like a capitalised NO!) but it's very effective and much more engaging style than a traditional, dry, engineering text.*

Not that many worked examples for an engineering textbook but, to be honest, the subjects are explained so well (with brilliant diagrams) that it is not really a problem.

Easily one of the best textbooks I have ever used. It was my 'secret weapon' during my studies and I'm convinced my good marks were almost entirely down to these two textbooks. I hope he writes more.

If you work with electrical equipment and systems, and want to know more about the science underlying the calculations you have to do, this is the best book to buy.

There are many other textbooks which concentrate on describing the various national and international electrical standards and wiring regulations in depth, but cover the basic underlying concepts of electrical science only very briefly.

This book is totally different and has been written to help anyone who really wants to understand how electricity works.

It is easy to dip into it wherever you wish, simply by using its list of contents and its comprehensive index.

However, if you want to learn in depth about electrical science, you can work through its chapters from cover to cover. If you can do that it would be like getting the instruction, coursework and hand-outs which the author has created over many years whilst he lectured about electrical engineering technology in naval and civilian colleges. He also worked as an instructional designer for various government and private training organizations.

In summary, this book is excellent. It deserves to become a standard textbook in use worldwide.

Having been involved in the training of electrical engineers and technicians over many years, I can thoroughly recommend this book for its clarity, conciseness and precision in both the written text and the illustrations (by the author). That the author has been closely involved in teaching this subject is obvious from the logical way the subject is approached and developed and from the regular opportunities given for revision and personal self-testing. I would give this book my strongest recommendation for any institution or individual engaged in the teaching or study of the principles of electrical engineering.

www.ingramcontent.com/pod-product-compliance
Lightning Source LLC
Chambersburg PA
CBHW062324220526
45469CB00008B/2618